Mission Command and the Grey Cell Protocols

Creating the Necessary Conditions for Decentralized Leadership

Kit Perez and Donald Vandergriff

Published by Calibration Press

An imprint of Rocky Mountain Arts and Music LLC

ISBN: 979-8-9952056-0-9

First edition, 2026

Printed in the United States of America

For permissions inquiries, bulk orders, or institutional licensing, contact the authors through their respective platforms: shepardscale.substack.com and donaldvandergriff.substack.com

To Eric, Alex, and the countless unnamed resisters of tyranny. — KP

To Lorraine, and my Reformer Friends in the Military and Law
Enforcement. — DV

FOREWORD

Mission Command is one of the most frequently cited and least understood concepts in modern military doctrine. Many organizations claim to practice it. Few actually create the conditions required for it to work.

Mission command is not simply decentralization or empowerment. It requires disciplined thinking, shared orientation, and a leadership culture that rewards competence, honesty, and initiative. Without those foundations, decentralization does not produce agility. It produces confusion and risk. The framework presented in this book recognizes that reality.

During combat operations in Afghanistan, the Marines of First Battalion, Sixth Marines faced similar challenges. Small units were often dispersed across difficult terrain and operating with incomplete information. Leaders at every level had to think independently while still acting within the commander's intent.

What made that possible was not simply training. It was the deliberate effort to build shared understanding and to develop leaders who could make sound decisions when isolated from higher command.

In practice, this required many of the same principles discussed in this book. We worked hard to create an environment where initiative was expected, where leaders understood the broader purpose of the mission, and where disciplined judgment was valued over blind adherence to procedure.

When those conditions exist, decentralized leadership becomes a powerful advantage. When they do not, it becomes a liability.

This book provides an important contribution to the discussion of how organizations can preserve the conditions necessary for mission command to function.

The Grey Cell Protocols offer a thoughtful framework for identifying and addressing many of the institutional weaknesses that undermine decentralized leadership. For military professionals, policy makers, and leaders in any complex organization, the lessons here are worth serious attention.

Mission command is not a slogan. It is a responsibility. And like any responsibility in war, it must be built deliberately.

For that reason, I highly recommend this work.

LtCol Asad Khan, USMC (Ret.)

PREFACE

This book began with a problem neither of us could solve alone.

Donald Vandergriff spent decades studying, teaching, and applying decentralized leadership (Mission Command) across military institutions, training centers, and deployed advisory environments. The pattern he kept encountering was consistent: organizations that understood the doctrine on paper, believed they were living it in practice, and failed in ways that surprised everyone except the people closest to the ground.

The failures followed predictable structural paths. The gap between what Mission Command required and what those organizations had built was the same, repeating across decades and domains. He could describe it precisely, but he did not have a system to close it.

Kit Perez developed the Grey Cell Protocols (GCP) from a different direction entirely. Working in civilian, volunteer, and resistance environments where formal authority is often absent, legal consequences are real, and infiltration is a live threat rather than a theoretical one, she kept encountering the same structural problem. Decentralized systems failed repeatedly, even with competent and loyal people, because the conditions that make decentralized command survivable had never been deliberately built. High-performing military units had always enforced those conditions, but through extreme selection, prolonged attrition, and institutional pressure that most organizations cannot sustain. GCP is the attempt to make that underlying logic explicit, transferable, and applicable in environments that cannot afford a permanent selection course.

When we began working together, the convergence was immediate. The structural gap Don had spent decades identifying in military institutions was the same gap GCP had been developed to fill. Mission Command named the destination; GCP was the infrastructure required to reach it. Neither framework was complete without the other, and neither of us had seen them treated as a single system.

This book is the result of that synthesis.

It is written for anyone who operates in a high-stakes environment that requires distributed authority, whether that environment is a military unit, a corporation, a church leadership team, an activist organization, or a resistance network.

The argument does not change across domains, and it is this simple: When the conditions Mission Command requires are absent, decentralized command fails. When those conditions are present, centralized control is unnecessary. The leader's job is to know which situation they are in and to have the tools to move between them deliberately.

That is what this book gives you.

Kit Perez
Donald Vandergriff
February 2026

AUTHORIAL SCOPE AND PERSPECTIVE

Donald Vandergriff brings decades of experience studying, teaching, and applying Mission Command and adaptive leadership within military and civilian institutions. His perspective is shaped by service across multiple roles, including uniformed leadership and deployed advisory work, giving him direct exposure to how command doctrine functions under real operational pressure and combat environments. *His research and writing focus on how command systems succeed or fail under uncertainty, particularly in environments where disciplined initiative, decentralized decision-making, and adaptability are required.*

Kit Perez is the developer of the Grey Cell Protocols, a systematic framework that makes explicit the conditions high-performing units have always enforced implicitly and extends them to environments that cannot rely on extreme selection or institutional attrition. She brings a background in applied behavioral analysis, counterintelligence methodology, trauma psychology, and group dynamics, with extensive experience in civilian, volunteer, and resistance environments. Her training includes graduate-level work in intelligence studies and trauma psychology, informing her analysis of orientation, deception, and human behavior under pressure. *Her work focuses on designing systems that stabilize orientation and make decentralized command viable under stress.*

INTRODUCTION: KABUL, 2021

In the summer of 2021, as U.S. forces prepared to leave Afghanistan after two decades of war, a company commander on the ground in Kabul watched events move faster than his chain of command could respond.

The situation was deteriorating by the hour. Taliban forces were advancing. Crowds were forming. Local partners who had worked with U.S. forces for years were suddenly exposed. Junior leaders and small units, positioned closest to the problem, could see what was coming. They requested authority to adjust security postures, accelerate evacuations, and respond to emerging threats before the window closed.

Those requests did not move quickly.

Decisions flowed upward instead of outward. Guidance returned slowly, filtered through multiple layers of staff coordination, approval gates, and risk-management checklists. Each update required slides, proposed adjustments demanded justification, and the decisions about whether to adapt came too slowly.

Opportunities to act passed while the process continued.

As control centralized, initiative narrowed. Units waited, their frustration and confusion increasing. The gap between what was happening on the ground and what was being briefed widened with each delayed decision.

When the Taliban entered Kabul with little resistance and the evacuation at Hamid Karzai International Airport devolved into chaos, the cost of that delay was no longer abstract. Lives were lost, and allies were left behind. The withdrawal was supposed to be an orderly retrograde, but it became a public and strategic failure.

Afterward, no single leader could be cleanly blamed. Procedures had been followed. Risk had been managed. Authority had been retained at the appropriate levels.

And yet, something had clearly gone wrong.

To understand *what* went wrong in Kabul, and why it goes wrong in the same way across military units, corporations, volunteer organizations, and resistance groups alike, requires a precise definition of a single concept.

Colonel John Boyd, the Air Force fighter pilot and strategist whose work on decision-making has shaped military doctrine for decades, gave us the framework.[1] Boyd's OODA loop, which stands for Observe, Orient, Decide, Act, is more than a decision cycle. It is a model for how humans and organizations stay connected to reality under uncertainty.

The loop reveals that advantage does not necessarily go to the strongest or the best-resourced, but to whoever can process change most fluidly. And the phase that governs that fluidity is orientation: the synthesis of new information with existing mental models, cultural traditions, previous experiences, and ongoing analysis to form a working map of reality.[2]

Orientation is often viewed as awareness, knowledge, or even merely perspective, but this is incorrect. Two people can observe the same event and arrive at completely different decisions because their orientation models differ. Orientation is the lens through which observation is interpreted, the filter that shapes what information gets weighted and what gets discarded, the framework that converts raw data into something actionable.

When orientation is healthy and continually recalibrated against reality, observation sharpens, decisions accelerate, and action becomes decisive. When orientation drifts, when the map a system uses to interpret the world no longer matches the terrain, the entire decision cycle degrades. Observation becomes selective, and decisions are slow or distorted. Action detaches from what the situation requires.

In a centralized system, distorted orientation is a leadership problem. In a decentralized system, distorted orientation is a command failure. When authority is distributed to subordinates who are interpreting reality through divergent or corrupted maps, decisions made under initiative will be wrong.

That is the mechanism behind every failure this book examines, and it is the problem the Grey Cell Protocols were designed to address at the source.

***

Mission Command, the warfare doctrine formalized by the U.S. Army, is a translation of the Prussian/German term *Auftragstaktik*, examined in depth in Chapter 2. The doctrine is often described as a leadership philosophy centered on trust, empowerment, and decentralization.

In civilian settings, it is commonly reduced to ideas such as delegating decisions, avoiding micromanagement, or giving professionals autonomy. When these approaches fail, the explanation usually focuses on people: they were not disciplined enough, mature enough, or aligned enough. In some cases, the failure is blamed on Mission Command itself as a "flawed" system.

This book makes a different argument; namely, that Mission Command is a conditional command system designed to function under uncertainty by distributing decision authority to the lowest competent level. It works only when specific human and structural conditions are already present. When those conditions are absent, Mission Command fails in predictable ways.

In many cases, it did not fail at all, because it was never instituted to begin with.

Leaders may have delegated authority to their subordinates, reduced oversight, or even adopted the language of trust and empowerment, but they never established the conditions that make decentralized command viable.

The result is assumed Mission Command, a false positive that produces drift, internal capture, burnout, and collapse. The results of that false positive show up differently depending on the environment.

- In a corporate environment, failure often manifests as stalled projects or high employee turnover.

- In a civilian resistance group, it can result in infiltration, entrapment, and arrests.
- For a military unit, the cost is paid in human lives.

The effects of assumed Mission Command result in leaders asking the same question across all domains:

Why does my organization keep failing in the same ways, even when I have good people?

The Central Claim

The central claim of this book is straightforward: Mission Command assumes its prerequisites; the Grey Cell Protocols build them.

Those prerequisites include shared orientation to reality, internalized discipline, emotional regulation under pressure, healthy embarrassment, and reliability in the absence of supervision. These conditions do not emerge naturally in most human systems. They must be deliberately designed, enforced, and protected.

The Grey Cell Protocols (GCP) exist to do exactly that work.

GCP is a command framework designed to create and maintain the conditions required for decentralized command in adversarial and high-stakes environments. Where Mission Command exploits alignment to generate adaptability and speed, GCP stabilizes alignment in the first place, allowing Mission Command to emerge.

The relationship between the two systems is a sequential loop rather than an interchangeable choice.

- GCP establishes orientation.
- Mission Command exploits orientation.
- Pressure degrades orientation.
- GCP is re-applied.

- Mission Command emerges once again.

This loop is continuous. There is no final state.

What This Book Is Not Arguing

This book does not claim that Mission Command is obsolete, broken, or in need of moral restoration. It does not argue that centralized control is superior. Nor does it suggest that civilian systems should imitate military organizations.

Our argument is narrower and more mechanical: Mission Command is the peak operational state of a high-trust, high-stakes system. It is unforgiving of misalignment with reality. Outside of environments that deliberately cultivate its prerequisites, it is frequently misapplied. When misapplied, its failures are blamed on people or doctrine rather than structure.

This book is a critique of assumptions.

Why This Applies Beyond the Military

Although Mission Command originates in military doctrine, the conditions it requires are systemic rather than unique to the military.

Any environment that involves the following facets faces the same control problem.

- distributed authority
- high trust
- time pressure
- ambiguous information
- real consequences

Church leadership teams, corporations, activist movements, volunteer organizations, and resistance groups all attempt decentralized execution. Across domains, failure occurs for the same reason: orientation is assumed rather than built.

How to Read the Chapters That Follow

Each section builds toward a single operational question that leaders must constantly face:

What is the current orientation state of the system, and which command framework is appropriate right now?

When orientation is unstable, Mission Command is unsafe. When orientation is stable, centralized control is unnecessary.

The leader's responsibility is not to prefer one doctrine over the other, but to **correctly diagnose the system state and apply the appropriate framework.** Failure most often occurs because the system was misread in practice.

What You Will Be Able to Do After This Book

Once you have read and understood this book, you will be able to do all of the following:

- Identify whether your system currently has the structural and psychological conditions required for decentralized leadership.
- Detect orientation drift before it becomes collapse.
- Distinguish between emotional trust and structural trust, and build the latter deliberately.
- Recognize early indicators of internal capture, narcissistic distortion, or incentive corruption.
- Diagnose when centralized control is temporarily necessary and when decentralized execution is safe.
- Design self-correcting systems that maintain discipline without constant supervision.
- Build teams capable of disciplined initiative under pressure rather than compliance under oversight.

- Evaluate whether your organization is operating under assumed Mission Command or actual Mission Command.
- Apply the Grey Cell Protocols to restore alignment when pressure, ego, or adversarial interference begins degrading cohesion.
- Create environments where initiative is safe because alignment is real.

Notes

1. John R. Boyd, "The Essence of Winning and Losing" (unpublished briefing, January 1996); John R. Boyd, "Destruction and Creation" (unpublished paper, September 3, 1976). Boyd's framework places orientation at the center of adaptive decision-making under uncertainty. For the most comprehensive secondary treatment of Boyd's strategic theory, see Frans P.B. Osinga, *Science, Strategy and War: The Strategic Theory of John Boyd* (New York: Routledge, 2007).

2. The formulation of orientation as "the synthesis of new information with existing mental models, cultural traditions, previous experiences, and ongoing analysis" follows Boyd's own language as summarized in Osinga, Science, Strategy and War, 165--178.

3. U.S. Department of the Army, *Mission Command: Command and Control of Army Forces*, Army Doctrine Publication (ADP) 6-0 (Washington, DC: Headquarters, Department of the Army, 2019).

PART I: THE SYSTEM AND THE PROBLEM

Mission Command is a conditional system state. GCP is the infrastructure required to create and maintain the conditions that Mission Command needs. Neither is complete without the other.

CHAPTER 1: THE CLAIM THIS BOOK DEFENDS

Mission Command fails when its required conditions are assumed instead of built.

In the summer of 1866, during the Austro-Prussian War, a Prussian corps commander received incomplete and conflicting reports about Austrian troop movements near Königgrätz. The reality on the ground was not a good one: the communications were unreliable, and the units were using outdated maps in poor visibility.

Rather than wait for clarification from higher headquarters, subordinate commanders acted on intent.

Helmuth von Moltke the Elder, serving as Chief of the Prussian General Staff, had issued broad objectives but avoided prescribing detailed execution. He understood that no plan would survive contact with the enemy and that commanders closest to the terrain would see opportunities long before headquarters could.

Several Prussian units advanced independently, exploiting gaps that had not appeared in earlier intelligence summaries. One division committed earlier than planned, another shifted axis without orders, and a third pressed an engagement that, on paper, appeared premature.

The execution was imperfect; there were errors, the formations were uneven, and their coordination left something to be desired, but their initiative was rewarded rather than punished.

Rather than intervening to reassert control, higher headquarters allowed the action to unfold. The Austrian forces, operating under more centralized command, were slower to respond. By the time their leadership recognized the developing threat, the tempo had shifted irreversibly.

What began as uncertainty resolved into a decisive advantage for the Prussians. Königgrätz became a turning point, because

Prussian leadership had trusted subordinates to act under uncertain conditions without waiting for permission.

This was the product of a command system that assumed uncertainty, tolerated visible error, and prioritized alignment with reality over adherence to a pre-determined plan.

That system would later be called *Auftragstaktik*. The U.S. military would later adopt its modern form as Mission Command.

What is often forgotten is that this form of command did not emerge naturally, and only functioned because specific conditions had been deliberately cultivated.

Mission Command Is Conditional, Not Aspirational

Mission Command only works under specific structural and psychological conditions. It cannot be adopted as a philosophy or declared into existence by leadership. It emerges only when orientation is structurally stabilized and the system has done the prerequisite work.

During the early phases of the 1940 German campaign in France, panzer units frequently operated beyond reliable communication with higher headquarters. Radio contact was intermittent, maps were incomplete, and front lines shifted faster than plans could be updated.

In several instances, subordinate commanders advanced on their own initiative, deviating from prescribed routes or timelines based on what they observed on the ground. These deviations were not always clean. Units often became temporarily misaligned while coordination suffered, leading to increased risk.

What did not happen is just as important.

Higher headquarters did not intervene to reassert detailed, centralized control. Subordinates were not retroactively punished for acting without permission, and visible error was tolerated in the service of battle tempo, because commanders shared a common operational orientation and the troops understood the commander's intent well enough to self-correct in motion. This worked because the system had already done the difficult work beforehand.

Officers were selected, developed, and socialized to think in terms of intent rather than instruction. Discipline was internalized rather than enforced through oversight. Initiative was expected, and healthy embarrassment from the error served as calibration rather than a career-ending event. Their structure made decentralized action survivable. Adaptability was already baked into their system, and decentralization enabled its exploitation.

This distinction between structural adaptability and informal autonomy matters because the same level of autonomy granted to a system without shared orientation, internal discipline, and tolerance for visible error would not have produced initiative but fragmentation, blame-seeking, and rapid recentralization.

This example is frequently misread as evidence that decentralization itself creates adaptability. Decentralization merely revealed capabilities that already existed. The command system did not manufacture judgment, discipline, or alignment under pressure; it assumed them. If those conditions were absent, the same autonomy would have produced failure rather than advantage.

Mission Command, therefore, is a conditional system state that emerges only when its prerequisites are deliberately engineered and continuously maintained.

Healthy Embarrassment as Calibration

Of all the prerequisites Mission Command requires, healthy embarrassment is the one most organizations are actively designed to eliminate.[3]

The concept is precise and worth defining carefully: Healthy embarrassment is the capacity to recognize your own error, accept correction from the system or from peers, and adjust your behavior without ego defense, external enforcement, or organizational punishment.

It is not the same thing as shame. Things like hazing and public humiliation are not healthy and will not have the same effect; in fact, when subjected to these things, people tend to react in an unhealthy way.

One of the first things the Prussians eliminated in their successful officer development system was hazing, which they found unprofessional. Instead, they put pressure and stress on their cadets and officers through problem-solving games under limited orders, time, and resources.

Embarrassment is the mechanism by which a decentralized system stays calibrated in real time.

In the Ardennes example, when a unit deviated from its prescribed route and created coordination problems, the system needed that deviation to surface visibly, be assessed honestly, and be corrected without the unit commander retreating into defensive posturing or the higher headquarters immediately centralizing to prevent future deviations. The error had to be visible and survivable. That is what healthy embarrassment enables: an environment where failure is information rather than a threat.

The reason this is counterintuitive is that most modern institutions have inverted the relationship between visible error and institutional response. Liability frameworks punish visible mistakes more harshly than hidden ones. Performance evaluation systems reward error avoidance over initiative. Informal culture treats visible failure as a signal of incompetence rather than a signal of honest engagement with a difficult problem.

The result is a system that teaches its people to manage the appearance of competence rather than develop actual competence. Errors get hidden rather than surfaced. Feedback gets softened rather than sharpened. Decisions get deferred upward because subordinates have learned that using authority and being visibly wrong afterward carries unacceptable career risk.

This is an incentive problem, and it is nearly universal.

A system without healthy embarrassment used as a fulcrum to learn from mistakes cannot run Mission Command. A feedback loop that allows a decentralized system to self-correct in motion depends entirely on errors being surfaced accurately and quickly. When that loop is severed, decentralization produces drift that accelerates in the absence of correction.

Chapter 4 examines healthy embarrassment as a structural condition and traces the specific mechanisms by which modern systems destroy it. The Grey Cell Protocols treat its presence or absence as a primary diagnostic indicator of whether a system is ready for decentralized command. What belongs here is the concept itself: the capacity to be visibly wrong without organizational consequence is the load-bearing condition that makes the feedback loop function.

Assumption is the Primary Failure Mode

Systems that claim Mission Command often collapse because its prerequisite conditions are presumed rather than built.

In a large emergency medical service operating across multiple jurisdictions, senior leadership publicly embraced decentralized decision-making. Supervisors were instructed to trust their crews, reduce micromanagement, and allow field units to exercise judgment under dynamic conditions.

On paper, the organization appeared aligned with Mission Command principles. In practice, it lacked a shared orientation about how the mission was to be carried out.

Protocols and constraints were based on the hospital's liability and billing practices rather than patient care. Liability constraints are legitimate operational parameters, and this organization had plenty of them. The failure was the absence of any stated mission that those boundaries were meant to serve. Crews had walls but no center. When pressure came, they had nothing to orient against except the constraints themselves.

Paramedics and EMTs had to constantly choose between what would help the patient in front of them and what policies they were told to follow. Training emphasized compliance over judgment. Performance evaluations rewarded error avoidance and documentation that minimized institutional risk rather than initiative or effective field care. Informal norms punished visible mistakes more harshly than delayed decisions.

Those who surfaced unresolved mismatches were put on undesirable shifts, counseled, or written up for insubordination. Crews learned quickly that deviation from standard procedure, even when tactically sound, carried career risk.

Over time, the gap between stated vision and actual incentives widened. Patient care quality dropped, and frustration among the crews on the ground mounted. Perfect charts and compliance ratings were touted as evidence of success.

A mass casualty incident revealed the truth.

Field units faced conflicting priorities in the moment: patient triage, scene safety, transport coordination, and interagency communication. Several crews made rapid decisions based on local conditions, intending to do the best they could for the largest number of patients.

When outcomes deviated from expectations, leadership could have responded by examining the orientation gaps and learning from them. Instead, after-action reviews focused on protocol adherence rather than situational judgment. Crews were verbally admonished for showing initiative and instructed to seek approval before deviating from standard procedures in future incidents.

The organization concluded that decentralization had been unsafe and moved back toward centralized micromanagement.

What had failed was the assumption that decentralized authority could function without shared orientation, internalized discipline, and tolerance for visible error. Delegation had occurred without the structural conditions that make initiative survivable. Mission Command had never been properly established.

This failure pattern repeats across domains because the assumption feels efficient. Building prerequisites is slow, uncomfortable, and politically costly. Assuming them allows leaders to claim empowerment without confronting the misalignment their own systems have produced.

Mission Command is unforgiving of this shortcut. When prerequisites are presumed rather than engineered, decentralization exposes system fragility. The resulting collapse is blamed on people, pressure, or the doctrine itself, rather than on the conditions that were never built.

That is the primary failure mode of assumed Mission Command, and it is the problem this book exists to solve.

Operational Closure

Mission Command is a conditional system state. Decentralization exposes whether adaptability already exists, but it cannot create adaptability. When Mission Command fails, it means that the prerequisites were never built. Healthy embarrassment is the load-bearing condition most organizations have structurally eliminated. Assumption is the primary failure mode.

What You Can Now Diagnose

- When an organization uses the language of trust and empowerment without building the structural conditions that make those things survivable.
- When visible error is punished more harshly than concealed error or deferred decisions.
- When decentralization has been declared but prerequisites have been assumed.
- When subordinates have authority on paper and protect themselves from using it in practice.
- When Mission Command is blamed for failures that were produced by its absence.

- When the feedback loop that allows a decentralized system to self-correct has been severed by incentives that punish visibility.

Chapter Notes

1.   For the Prussian development of Auftragstaktik and its institutionalization under Moltke the Elder, see Gordon A. Craig, The Battle of Königgrätz: Prussia's Victory over Austria, 1866 (Philadelphia: University of Pennsylvania Press, 2003), 112--145; Robert M. Citino, *The German Way of War: From the Thirty Years' War to the Third Reich* (Lawrence: University Press of Kansas, 2005), 131--168. For the U.S. Army's adoption of Mission Command as modern Auftragstaktik, see ADP 6-0 (2019).

2.   The Ardennes campaign material draws on Karl-Heinz Frieser, *The Blitzkrieg Legend: The 1940 Campaign in the West*, trans. John T. Greenwood (Annapolis, MD: Naval Institute Press, 2005), 204--248.

3.   The concept of healthy embarrassment as a specific structural calibration mechanism is developed in this book in collaboration. For Vandergriff's prior work on calibration and feedback in military organizations, see Donald E. Vandergriff, *Adopting Mission Command: Developing Leaders for a Superior Command Culture* (Annapolis, MD: Naval Institute Press, 2019).

CHAPTER 2: WHAT MISSION COMMAND IS

Mission Command is a decentralized command doctrine designed to generate adaptability and disciplined initiative in uncertain conditions.

In May 1940, German armored forces pushed through the Ardennes forest, terrain the French high command believed was unsuitable for large, mechanized formations. The plan was audacious, with incomplete intelligence and unreliable communications.

What followed would redefine modern warfare.

General Heinz Guderian's XIX Panzer Corps was ordered to cross the Meuse River near Sedan and exploit westward toward the English Channel, nearly two hundred miles deep into enemy territory. Higher headquarters issued mission-type orders with a simple intent: reach the coast with all possible speed.

When French resistance stiffened along the Meuse, Guderian did not pause to request detailed guidance. On his own authority, he massed artillery and close air support, forced the crossing, and immediately pushed armored spearheads forward. This exposed his flanks, and communications with higher-ups were intermittent.

As the advance accelerated, the German high command grew nervous. Reports warned that the panzer columns were outrunning infantry support and leaving gaps that could be exploited. Suggestions were sent forward to slow the advance, consolidate gains, and restore control.

Guderian and other corps commanders, including the new commander of the 7th Panzer Division, Erwin Rommel, pressed on, ignoring their orders.

They correctly judged that Allied disarray mattered more than textbook caution. Subordinate commanders bypassed resistance, exploited gaps as they appeared, and adjusted routes without waiting for approval. Decisions were made at the point of contact by leaders who could see the situation on the ground unfolding in real time.

On May 20, German armor reached the Channel coast.

The British Expeditionary Force and much of the French Army were cut off. What had begun as a risky penetration became a decisive operational collapse in spite of the caution urged by high command, because authority moved with the information, not the rank.

The operation was far from clean, and many errors were made. What mattered was that initiative was rewarded rather than punished, and that visible error was tolerated in the service of tempo.

Command, Not Leadership

Mission Command governs a specific and bounded set of questions: who decides, when they decide, and under what constraints. It does not govern personality, motivation, or management style.

This distinction is important because most institutions that claim Mission Command are attempting to solve a leadership problem with a command doctrine, and the two are not the same thing.

When the definition of Mission Command is left ambiguous, personality fills the gap. In the absence of clear authority boundaries, the system defaults to whoever is most willing to act, most politically connected, or most risk-tolerant. The result looks like decentralization but functions as informal centralization, because real authority never moved from wherever it was most comfortable.

Without shared meaning grounded in observable behavior, Mission Command cannot be practiced, only discussed.[2]

The evidence for this is operational. During Don's advisory work in Kabul from 2013 to 2018 with Resolute Support US/NATO Command, the same failure appeared repeatedly across coalition and Afghan units alike. American advisors spoke confidently about empowering Afghan partners, trusting local commanders, and encouraging initiative. Afghan leaders echoed the same language. In briefing rooms and planning sessions, driven by PowerPoint briefings made by hundreds of staffers and contractors, the doctrine appeared shared.

In execution, it was not.

When pressure increased, advisors tightened oversight. Afghan commanders hesitated, waiting for approval even when circumstances clearly demanded action. Both sides believed the other was failing to live up to Mission Command principles, but neither could precisely identify where the breakdown was occurring.

The breakdown was definitional. Each group was using the same words to describe fundamentally different command behaviors. To one side, trust meant tolerance of initiative; to the other, it meant insulation from accountability. Initiative meant adaptation to some and permission-seeking to others. The definition of command itself ranged from intent-setting to detailed control, depending on who was speaking and under what pressure.

The system was failing because the foundational questions had never been answered with precision: who decides, when they decide, and what constraints govern the decision space.

Until those questions are answered and the answers are shared and enforced, discussions about restoring Mission Command produce noise rather than outcomes.

Decentralization as a Control Strategy

The most common misreading of Mission Command treats decentralization as an act of trust in subordinates. It is more precise than that. Decentralization is a control strategy, chosen because it is the only way to maintain control over outcomes when conditions change faster than headquarters can process and respond to them.

In high-tempo environments, centralized control does not provide safety. It provides latency. By the time a decision travels upward for approval and returns with guidance, the conditions that generated the decision have often changed. The guidance arrives calibrated to a situation that no longer exists. The gap between headquarters understanding and ground reality widens with each iteration, and the system loses the tempo it was designed to generate.

Decentralization solves this by repositioning control authority at the point where information is freshest and time pressure is highest. It is not a reduction of control. It is a redistribution of control to where it can function.

Don's observation of force-on-force rotations at the National Training Center produced a consistent and instructive pattern. Units that treated decentralization as empowerment struggled. Units that treated it as a control mechanism adapted faster and survived longer under pressure.[3]

In one rotation, a battalion commander entered a fast-moving fight against a thinking opposing force. He issued a short intent-based order focused on disrupting enemy command and isolating maneuver elements. He did not issue detailed movement plans beyond initial boundaries and constraints.

As contact developed, company commanders encountered conditions that diverged rapidly from the planning assumptions. One company identified an exposed enemy flank that was not part of the original scheme of maneuver. Another encountered stronger resistance than anticipated and risked becoming decisively engaged. Neither waited for permission to act.

One company shifted its axis to exploit the flank. The other broke contact early and repositioned to preserve combat power and support the battalion's larger objective. Both actions deviated from the original plan.

From outside the command system, the execution looked disorderly. From inside it, the actions were tightly controlled. The battalion commander did not intervene to correct execution details. He adjusted fires, shifted resources, and refined end-state guidance as reports came in. Control was exercised through intent and resource allocation, not through directing movement.

The opposing force, operating under more centralized control, reacted more slowly. Their decision cycle required confirmation, synchronization, and approval at each stage. By the time adjustments were authorized, the battalion had already changed the conditions on the ground.

The after-action review did not examine whether companies had followed the plan, but whether actions aligned with intent, whether risks were taken prudently, and whether decisions were fast enough to outpace the opponent.

The battalion's success was a product of control positioned where information was freshest and time mattered most. When decentralization is treated as permission rather than control architecture, leaders feel exposed and unsafe. They respond by recentralizing decisions, increasing coordination requirements, and slowing the system. The doctrine gets blamed for a failure that was produced by its misapplication.

Why It Is Often Reduced to Slogans

The language of Mission Command (often referred to simply as decentralized leadership) spreads easily across civilian and institutional environments because it sounds like what every organization already believes it is doing. Trust your people. Push decisions down. Reduce micromanagement. These are not wrong ideas, but they are not Mission Command. They are its surface features, detached from the structural conditions that make those features functional.

What gets discarded in translation is the discipline: the precise definition of intent, the non-negotiable constraints, the tolerance for visible error, and the enforcement of shared orientation through feedback and healthy embarrassment. Without those elements, decentralization exposes whatever misalignment the system was already carrying.

Chapter 5 examines this failure pattern in detail. What belongs here is the observation that Mission Command is consistently misread as a cultural value rather than a command architecture, and that the misreading produces a predictable outcome. The slogan survives because it is aspirational. The discipline is avoided because it is uncomfortable. Under pressure, the system behaves exactly as it was built to behave.

Operational Closure

Mission Command governs decision authority, not personality or motivation. Decentralization is a control strategy chosen because it is the only mechanism that keeps pace with conditions that change faster than headquarters can process.

Control shifts to the point of contact because that is where information is freshest and time pressure is highest. Intent, defined constraints, and tolerance for visible error are critical elements.

When those definitions are absent or ambiguous, personality fills the gap. The system defaults to informal centralization, and the doctrine takes the blame for conditions it was never given the infrastructure to address.

What You Can Now Diagnose

- When an organization uses the language of trust and decentralization without defining who decides, when they decide, and under what constraints.
- When "initiative" and "trust" mean different things to different people in the same system, and the difference has never been surfaced or resolved.
- When decentralization was announced rather than built, the structural conditions were never established.

- When after-action reviews focus on whether the plan was followed rather than whether actions aligned with intent.
- When control has been re-centralized informally through social enforcement rather than through explicit command authority.
- When Mission Command language is present throughout the organization, yet Mission Command outcomes are absent.

Chapter Notes

1. For the Ardennes breakthrough and Guderian's conduct at the Meuse crossing, see Karl-Heinz Frieser, *The Blitzkrieg Legend: The 1940 Campaign in the West*, trans. John T. Greenwood (Annapolis, MD: Naval Institute Press, 2005), 204--248.

2. The Kabul advisory material draws on Donald E. Vandergriff's contractor service with DynCorp International supporting NATO Resolute Support Mission, Kabul, Afghanistan, 2013--2018.

3. NTC rotation observations draw on Vandergriff's training center work.

CHAPTER 3: THE GAP AND THE SYSTEM THAT FILLS IT

Mission Command cannot create its own preconditions. The Grey Cell Protocols exist to do that work.

During a combat training center rotation, Don observed a U.S. battalion that had been explicitly instructed to operate under Mission Command principles. Higher headquarters issued clear intent, reduced reporting requirements, and deliberately limited interference during execution. On paper, the conditions for decentralized execution were in place.

In practice, company commanders repeatedly deferred decisions that fell well within their authority. Platoon leaders waited for confirmation before exploiting opportunities that were obvious in real time. Tactical pauses emerged not because leaders hesitated to act without additional reassurance.

After the rotation, senior observers initially attributed the failure to individual caution or lack of confidence. That explanation did not survive scrutiny.

During after-action reviews, a consistent pattern emerged. Leaders understood the intent and knew they were authorized to act. What they did not trust was the system's response after the fact.

Several officers described prior experiences outside that rotation where acting decisively had resulted in investigations, career damage, or informal labeling as "reckless," even when outcomes were positive overall. Others described watching peers succeed tactically while suffering professionally because results did not align with preferred narratives or metrics.

Before the rotation had even begun, the participants had learned the lessons.

Under pressure, leaders calculated institutional risk instead of tactical or even strategic risk. The safest courses of action were delay, coordination, and permission-seeking, even though the doctrine specifically called for the opposite, all because experience had taught them which behaviors were survivable. Mission Command was present as a written and taught doctrine, but as a lived reality, it was nowhere to be found.

Delegating authority did not restore initiative because the system had never protected it. There was no reliable tolerance for visible error, no consistent calibration mechanism, and no insulation against retroactive punishment. Their supposed autonomy did not correct misalignment.

When authority was pushed downward into an environment that had already been shaped by fear of consequence rather than trust in correction, decentralization amplified hesitation instead of adaptability.

Mission Command failed as a condition generator, even for the very conditions it needs to emerge.

It assumed alignment, discipline, and psychological safety for initiative, but possessed no internal mechanism to create them. Without those preconditions, decentralization was dangerous instead of empowering and adaptable.

This pattern repeats wherever Mission Command is introduced without first rebuilding the conditions that allow it to function. The doctrine can exploit alignment, but it cannot manufacture it. That structural gap is the problem this chapter examines.

Autonomy Does Not Correct Misorientation

Distributing authority into a misaligned system does not correct the misalignment. It amplifies whatever is already there.

In one cell-based resistance network operating without a formal hierarchy, leaders deliberately attempted to adopt Mission Command principles. They had grown disillusioned with other groups that relied on tight central control, personality-driven leadership, and constant approval cycles. Decentralization looked like the solution to problems they had watched destroy other organizations.

The network was composed of capable, motivated people with shared goals and a strong moral framework about what they were doing. Many believed that removing hierarchy would unlock initiative and prevent the abuses they had seen elsewhere. Formal leadership roles were intentionally dissolved. Cells were encouraged to act independently in pursuit of broad objectives, each operating as a self-directed node.

What the group had not addressed was that two critical prerequisites were missing: shared orientation and clear constraints, which could have been outlined in a Mission Command-aligned methodology. In their absence, decentralization produced divergence and eventually collapse.

Some cells interpreted autonomy as permission to escalate aggressively and publicly. Their actions drew attention, damaged the operating environment, and increased the risk for other cells working quietly in the same area. Several members were arrested. Inside those same cells, members who disagreed with the escalation had no safe mechanism for challenging it. Many of them simply walked away from both the cell and the broader effort.

Other cells reacted in the opposite direction. With no clear boundaries, decision criteria, or shared model for acting under uncertainty, freedom turned into paralysis. Members broke operational security by reaching out to others for guidance, trying to determine what was allowed. Taking initiative felt risky because there was no way to know how an action would be judged after the fact. In the absence of orientation, autonomy became inertia. These cells drifted into online signaling and meme culture, which created the feeling of participation without the risk of action.

A third pattern emerged simultaneously. Some members experienced the absence of visible leadership as instability rather than trust. To resolve their discomfort, they informally installed a leader within the cell, often themselves or someone they trusted emotionally. Authority returned, but through anxiety rather than structure. Control became personal, inconsistent, and unaccountable.

None of these outcomes was driven by bad intent. They were predictable results of distributing authority into a system that had not established shared orientation, internal discipline, or tolerance for visible error. Autonomy amplified what was already missing.

Mission Command was never truly present. Authority had been distributed into an environment where orientation was unstable, incentives were social rather than corrective, and error carried identity-level consequences. Under those conditions, freedom was the catalyst for already-present failure modes.

Decentralization Increases Risk When Orientation Is Unstable

The same mechanism that produced collapse in the resistance network appears in formal institutional environments, with different enforcement tools and equally predictable results.

During coalition advisory efforts in Afghanistan, advisory teams were frequently given significant latitude to adapt locally.[1] On paper, this looked like decentralized command. Junior advisors were encouraged to tailor their engagement with Afghan units, adjust approaches based on local conditions, and exercise judgment rather than wait for detailed guidance from above.

What did not change was the surrounding system.

Metrics remained centralized and reporting requirements still rewarded positive narratives over accurate assessments. Advisors were evaluated less on whether their actions improved Afghan unit capability and more on whether they avoided controversy, produced reassuring reports, and aligned with headquarters' expectations. Honest failure carried professional and political cost.

Some advisors responded to the newfound autonomy by taking aggressive, unilateral action. They pushed partner units into operations that looked decisive on paper but were misaligned with local realities. In several cases this meant overstating Afghan readiness, escalating operations for short-term gains, or prioritizing visible success over sustainable outcomes. The autonomy amplified misjudgment rather than correcting it.

Other advisors moved in the opposite direction. Aware that visible mistakes would not be protected, they avoided initiative altogether. Decisions were deferred upward and local adaptation stalled. Authority technically existed, but exercising it felt unsafe. Waiting became the rational choice.

In both cases, decentralization increased risk.

What appeared from the outside as inconsistent leadership or uneven advisor quality was a structural problem. Authority had been distributed into a system that punished honest feedback, blurred acceptable risk, and provided no reliable mechanism for recalibration when decisions went wrong. The result was not Mission Command, even though every party involved believed that was the model they were using.

When orientation is unstable, decentralization distributes risk. It does not generate the adaptability it is designed to produce.

The Structural Gap GCP Was Developed to Fill

The failures described in the previous sections are consistently misunderstood as leadership problems, cultural problems, or individual training shortfalls. They are in fact structural problems, and they point to a specific and identifiable gap.

Mission Command is an exploitation doctrine. It describes how authority should be used once alignment already exists. It relies on shared orientation to reality, internalized discipline, clearly understood boundaries, healthy embarrassment as a recalibration mechanism, tolerance for visible error, and protected channels for early correction. When those conditions are present, decentralization produces speed, adaptability, and disciplined initiative. When they are absent, decentralization magnifies misalignment and accelerates collapse.

Mission Command does not explain how to build those conditions, how to stabilize them under pressure, or how to repair them when they degrade. The doctrine was built to exploit alignment, not to create it.

That missing work is the structural gap. Using Mission Command to fix misorientation is like using speed to correct steering failure. It feels decisive but significantly increases the danger.[2]

The Grey Cell Protocols exist specifically to close this gap.

GCP is the infrastructure layer that does the work Mission Command cannot do by itself.[3] It stabilizes orientation before authority is distributed, defining non-negotiable constraints so that initiative has shape rather than just freedom. It creates protected mechanisms for surfacing errors early, before they propagate into failure. It enforces discipline without relying on constant oversight. It prevents identity, ego, and narrative from substituting for reality in the system's decision inputs.

Mission Command asks: who should decide, and how much freedom do they have to act?

GCP asks: Is this system oriented enough to survive that freedom?

Together, the two frameworks form a complete command doctrine. GCP builds and protects alignment. Mission Command exploits alignment for adaptability and tempo. Used in a continuously recalibrating sequence, they explain not only how high-trust systems succeed but why so many fail while believing they are doing everything right.

Why Elite Units Are Not a Counterexample

A common objection emerges here. If Mission Command cannot generate its own prerequisites, how do elite units such as Navy SEALs, Special Forces, and similarly high-performing teams operate so effectively without the Grey Cell Protocols?

The short answer is that they are operating inside systems that already enforce the principles that GCP formalizes, at extraordinary cost.

Elite units are sometimes cited as proof that Mission Command functions when leaders trust their people enough. These units succeed precisely because they exist inside environments where alignment, discipline, and correction are assumed only after they have been ruthlessly cultivated. Selection, sustained attrition, and continuous pressure are the mechanisms by which the prerequisites are built and maintained.

Elite units rely on extreme entry standards, prolonged and repeated assessment, relentless internal correction, and constant exposure to failure under supervision. Most people never pass these filters. Some who do pass initially are removed later, or self-select out. The cost in time, resources, and human attrition is enormous.

What remains is a population that has already internalized shared judgment, tolerates correction without ego defense, and can operate under uncertainty without collapsing into paralysis or narrative protection. The system was designed to eject those who cannot or will not internalize the discipline necessary to function within it. Mission Command works there because the prerequisite work has already been done.

What elite units demonstrate is not that Mission Command can function without prerequisites. They demonstrate that when prerequisites are enforced aggressively enough, Mission Command becomes extraordinarily powerful.

The limitation is that this model does not scale.

Most military units, civilian organizations, volunteer systems, churches, corporations, and resistance groups cannot afford multi-year selection pipelines, absorb extreme attrition, or rely on constant interpersonal pressure to maintain alignment. When these environments attempt to replicate elite-unit autonomy without replicating the invisible conditioning beneath it, they reproduce only the surface features: freedom without discipline, trust without correction, decentralization without shared orientation.

The result is collapse.

The question is how to deliberately create the conditions that those units rely on in environments that cannot function as a permanent selection course. GCP is the answer to that question.

The Grey Cell Protocols: A System Map

GCP is a five-facet command infrastructure. Each facet addresses a different layer of the same problem: how to design a system that remains oriented to reality under pressure, survives internal and external threats to that orientation, and maintains the conditions Mission Command requires without depending on institutional attrition or extreme selection.[3]

The five facets operate in sequence and in parallel. They are not independent tools, and cannot be applied singly. They are a system, and the system crosswalk is this:

GCP I establishes self-disqualifying systems. The environment is designed so that misaligned actors expose themselves and exit rather than waiting for the organization to identify and remove them. The burden of vetting shifts from leadership vigilance to environmental design. Culture becomes the filter.

GCP II builds structural trust. The system is architected to survive betrayal, error, and failure without experiencing them as collapse. Emotional trust, which depends on personal bonds and is therefore fragile under pressure, is replaced by structural trust, which depends on design and is therefore durable under pressure.

GCP III applies psycholinguistics. Language is trace evidence of orientation. Trauma explains motive. Together, they produce an early warning system that detects orientation drift before it manifests as behavioral failure. The system reads what people say and how they say it as a diagnostic mechanism.

GCP IV maps infiltration archetypes. Infiltration is predictable because human psychology does not change. Infiltrators and manipulators repeat the same playbook regardless of domain. Naming the archetype shifts the group from gut instinct to pattern recognition, and from suspicion to diagnosis. Detection becomes systematic rather than interpersonal.

GCP V establishes counter-narrative control. Narrative is the attack surface of any orientation-based system. Counter-narrative control keeps the story aligned to reality so that the rest of GCP does not ingest corrupted inputs. It is the meaning-integrity layer that sits on top of the other four facets.

The relationship between GCP and Mission Command is a sequential loop:

GCP establishes orientation. Mission Command exploits orientation. Pressure degrades orientation. GCP is re-applied. Mission Command emerges again.

This loop is continuous, constantly in a recalibrating state. Part II of this book examines why the conditions break down and what failure looks like when they do. Part III delivers the operational doctrine for each facet in full.

The reader now holds the complete architecture. Everything that follows is evidence for why each facet is necessary and instruction for how each facet works.

Operational Closure

Mission Command assumes alignment but cannot create it. Autonomy amplifies whatever orientation already exists in the system. When orientation is unstable, decentralization increases risk rather than generating adaptability. Initiative cannot survive where error is punished retroactively, and the system's response after the fact is unpredictable.

Mission Command is an exploitation doctrine. GCP is the infrastructure that makes exploitation safe.

Elite units are not a counterexample. They demonstrate the power of prerequisites that have been ruthlessly enforced, at a cost most organizations cannot sustain. GCP makes those prerequisites transferable.

What You Can Now Diagnose

- Whether decentralization in your system will produce adaptability or fragmentation depends entirely on whether the prerequisite conditions are present before authority is distributed.
- Whether fear of institutional consequence is suppressing initiative in your system, and whether that fear is a character problem or a structural one.
- Whether your metrics, incentives, or reputation management systems are stabilizing or destabilizing orientation.
- Whether your system has any protected mechanism for early correction, and what happens to people who use it.
- Whether performance in your environment is built on scalable structure or on individual attrition and informal pressure.

The diagnostic question that governs everything that follows is, "If you pushed authority downward in your system today, what exactly would it magnify?"

Chapter Notes

1.　Afghanistan advisory material draws on Donald E. Vandergriff's contractor service with DynCorp International supporting NATO Resolute Support Mission, Kabul, Afghanistan, Jan 2013-Jul 2018.

2. For more on the pop-OODA concept that speed is the goal of the OODA loop, see Perez's article "John Boyd Was Not Your Productivity Guru," Substack, July 22, 2025. https://shepardscale.substack.com/p/john-boyd-was-not-your-productivity

3. See Perez, "Orientation is the System," Substack, November 9, 2025. https://shepardscale.substack.com/p/orientation-is-the-system>

PART II: THE CONDITIONS AND THEIR ENEMIES

The conditions Mission Command requires do not emerge naturally, and modern systems predictably destroy them. The failure patterns that result are domain-independent and recognizable before they become a collapse.

CHAPTER 4: WHAT MISSION COMMAND REQUIRES AND HOW SYSTEMS DESTROY IT

The conditions that Mission Command requires are observable system properties. When they are absent, decentralization becomes unsafe rather than adaptive.

In Afghanistan in 2003, an Army infantry unit was going out on patrol. It was a route they had done countless times before, and the Taliban always reacted the same way. As the soldiers would make their way down the trail, the Taliban fighters would shoot a few rounds over their heads. Comparatively speaking, there was no real danger; neither side wanted an actual firefight. Two or three rounds would fly well over their heads, and that would be it.

This time, however, was different. A brigadier general would be accompanying the unit on their rounds. He had heard about the "easy" patrol route that never ended in casualties and saw it as an opportunity for him to get a Combat Infantryman's Badge with little risk. Armed with nothing but a sidearm, he announced that he would be coming along.

Every man in the unit knew what he was doing, and no one respected it. They went on their patrol, the expected shots came, and the general got his badge.

Other officers heard about it, and soon the unit saw several colonels also show up for their "CIB patrol" as well. This caused massive discontent in the unit.

The sticking point, as it were, was not that the men resented having senior officers with them. It was the blatant use of their lives for personal gain without any kind of embarrassment for having done so, even while looking them in the face.

More than twenty years later, the men still discuss that period with a level of mocking derision that reveals how deeply it affected them at the time.[1]

That incident not only insulted every man in the unit, but it destroyed the conditions Mission Command requires. The sections that follow explain how, and what it looks like when each condition is absent.

Shared Orientation to Reality

What made the 2003 patrol incident corrosive was the fracture of shared reality.

Every soldier on that patrol understood the actual situation on the ground. They knew the route was predictably low-risk. They knew the Taliban's behavior was ritualized rather than lethal. And they knew exactly why the brigadier general had chosen that particular patrol.

His behavior signaled something unmistakable: the official narrative and the lived reality were no longer the same thing. On paper, the patrol could be described as combat exposure. In his case, it was a controlled performance designed to produce a credential. All parties knew it. And critically, everyone also knew that no one would acknowledge it openly for fear of repercussion.

That unspoken split is what orientation failure looks like at the unit level.

Shared orientation means that actors across a system perceive the same terrain, the same risks, the same incentives, and the same truths, even when those truths are uncomfortable. It is the condition that allows intent to be interpreted consistently and initiative to be exercised safely. In this case, the general and the unit were operating from entirely different maps. The soldiers' map reflected ground reality: predictable contact, symbolic risk, and instrumentalized exposure. The general's map reflected status incentives: credential acquisition at minimal personal cost. The general's rank allowed the behavior to pass without correction.

Once that divergence occurred, Mission Command was already impossible regardless of how much authority might later be delegated.

Decentralized decision-making assumes that when intent is issued, subordinates interpret it against the same reality the issuer believes they are operating within. When reality fractures, intent becomes ambiguous and initiative becomes dangerous. The unit's sustained mockery and resentment were not emotional fragility but an accurate diagnostic signal. The system had revealed that truth was subordinate to optics, and that senior leaders could exploit risk and even their own men asymmetrically without consequence or embarrassment.

From that moment forward, any attempt to invoke trust, professionalism, or shared purpose would ring hollow. The soldiers had learned what the system taught: that what is said officially does not govern outcomes, that feedback about blatant disrespect is not safe to offer, and that their existence could be used to facilitate the advancement of someone whose career carried more institutional value than their lives.

Shared orientation cannot be assumed. It requires a common understanding of actual risk rather than rhetorical risk, alignment on what behaviors are genuinely rewarded rather than what is publicly praised, agreement about what will be acknowledged openly versus quietly ignored, and confidence that reality, when surfaced, will not be punished. Without any one of these, decentralization exposes actors to asymmetric consequences, where those closest to danger bear the cost while those above them extract the benefit.

Internalized Discipline

During a high-tempo training rotation at the National Training Center, Don witnessed a battalion operating under genuine Mission Command providing the clearest illustration of what internalized discipline looks like in practice.[2] The commander issued a concise intent focused on disrupting enemy command and isolating maneuver elements. Boundaries were clear. Execution authority was pushed to the units on the ground.

Company commanders encountered opportunities that tempted overextension. Platoons found exposed flanks they could exploit aggressively. Junior leaders could have pressed forward for personal distinction or short-term tactical gains.

Instead, they exercised restraint. Units disengaged early when risk exceeded the intent. Leaders abandoned tactically attractive options that still violated the stated constraints. Decisions were made quickly but deliberately, always weighted against intent and parameters.

No one stopped them or corrected them in real time. That restraint was already structurally present.

Internalized discipline is the condition that allows authority to move downward without the system fragmenting. It is more than integrity, though integrity is a component of it. It is consistency between stated values and observed behaviors under freedom, not just under oversight.

A person with integrity will do what they believe is right even when it is costly. That is admirable, but insufficient for Mission Command. The type of discipline Mission Command requires goes further: individuals reliably regulate their own behavior in alignment with intent and constraints without external enforcement, and that regulation is visible enough over time for others to observe and confirm it. That observability is what builds structural trust.

Mission Command assumes this condition already exists at the structural and personnel level.

The contrast is visible in systems where discipline is external. Checklists, approval chains, reporting requirements, and the threat of retroactive punishment can produce orderly-looking behavior. These systems collapse under speed and pressure. By the time correction arrives, the damage is already done. And more critically, under those conditions, the safest behaviors are delay and permission-seeking, which are precisely what Mission Command is designed to eliminate.

Discipline cannot be injected at the moment of decision; it must already exist. When authority is distributed into a system where discipline is external rather than internalized, one of two outcomes follows: actors overreach, chasing recognition, reward, or the momentum of the moment, or actors freeze, unwilling to act without explicit cover. Both are rational responses to an environment that has never required or rewarded self-governance. Mission Command treats internalized discipline as fuel; it consumes it but cannot produce it.

Healthy Embarrassment and Error Correction

Professional military cultures once treated embarrassment as a vital tool for alignment, and the mechanisms they built around it are worth examining before describing their absence.

In the Prussian General Staff system under Moltke the Elder, post-battle critiques were rigorous, expected, and structurally protected.[3] Officers documented failures without euphemism. A commander who misread enemy disposition or failed to exploit an opportunity would stand before peers and superiors, acknowledge the misalignment, and explain how he would adjust. The critique was sharp, professional, and impersonal: not an attack on character, but a calibration of the map against the terrain. The early Reichswehr carried this forward. In the interwar period, officers who failed to adapt during exercises faced pointed questions from peers, with clear exposure of flawed assumptions and no personal attacks. The process built trust because subordinates knew leaders would calibrate honestly, and leaders knew the process would improve collective orientation.

Retired Navy SEAL Cory Zillig spoke candidly about the error correction mechanism at the team level.[4]

> *It hurt. It was uncomfortable. It takes bravery to dish out harsh words meant for good. And it takes more than bravery to take them; humility is heavily involved. And none of it was anonymous!*

This is what healthy embarrassment as a functional system looks like: the constructive discomfort that signals misalignment between expectation and reality, prompting recalibration before catastrophe. It is not shame, which is global and personal, attacking identity rather than behavior. It is not hazing, which is ritualized humiliation unrelated to performance.

Healthy embarrassment is specific to the misalignment, behavioral rather than personal, temporary, proportionate, and forward-looking. It is the pilot's instrument warning. When the artificial horizon shows a bank but the physical sense says level, a professional pilot trusts the instruments, corrects, and realigns. That momentary discomfort saves lives. The same mechanism, operating in a command system, allows errors to surface and be corrected before they compound.[5]

The brigadier general's CIB patrol was not a violation of any written procedure. No constraint existed that would have prevented it. But it was a failure of orientation and restraint, and it should have triggered embarrassment. It did not.

Yet the general did not acknowledge the asymmetry of risk. He did not receive correction from peers, certainly not from the men whose lives he was using so callously, and no correction was imposed from his superiors. The behavior passed without comment, explanation, or consequence because the system allowed nothing else.

Once other senior officers followed suit, the message to the unit was unmistakable: personal advancement could be extracted from subordinate risk without shame, correction, or accountability. From that point forward, error correction inside the system was no longer possible. The soldiers could see the misalignment clearly, but they had no protected mechanism for surfacing it. Doing so would have required accusing senior officers of exploitation, an action that carried career and social consequences far greater than any benefit.

The soldiers' long-term mockery and derision were not a failure of decorum or morale. Mockery is what fills the vacuum when correction is structurally impossible. It was evidence that the internal corrective mechanism was already gone.

In a system capable of Mission Command, the incident would have ended differently. The behavior would have been recognized as misaligned. Embarrassment would have triggered restraint or prevented the incident entirely. Peers or superiors would have corrected it openly and proportionately. Trust would have been preserved through acknowledgment rather than silence. Instead, the system insulated the offender and exposed the subordinates. That inversion permanently damaged shared orientation and the unit's structural trust.

Decentralized command requires leaders who can absorb embarrassment without ego collapse, and subordinates who trust that surfacing errors will not destroy them. When that condition is absent, error correction dies, and with it, Mission Command.

How modern systems destroy it

The mechanisms that killed healthy embarrassment in that unit are not unique to that unit. They operate systemically, and they have institutional names.

Zero-defect culture is the most documented.[6] When promotion systems reward flawless records and visible mistakes trigger investigations rather than learning, leaders stop placing themselves in situations where they might be visibly wrong. They centralize decisions to minimize exposure; they will sanitize reports, focus on metrics over results, and avoid honest after-action reviews. The rational incentive is to manage the appearance of competence rather than develop actual competence, and the system trains that behavior reliably.

Legal and risk-management frameworks amplify the effect. Exhaustive documentation requirements, approval chains, and the possibility that candid self-critique will become evidence in an adverse action make honest acknowledgment of misjudgment professionally dangerous. Better to sanitize. Better to attribute failure to circumstances rather than to one's own flawed map of reality.

A third mechanism operates at a deeper layer. Broader cultural shifts have reframed constructive discomfort as potential harm. What was once professional rigor is increasingly pathologized as toxic feedback or psychological danger. Leadership training often prioritizes emotional safety over candid critique, after-action reviews get softened with appreciative inquiry framing, and feedback processes with pre-briefs to avoid discomfort. All of these eliminate the calibration signal while preserving the appearance of a learning culture. The intent is often compassion but the effect is insulation from reality. Orientation drifts because the mechanism that corrects it has been quietly removed.[7]

These accelerators interlock. Compliance culture avoids legal risk. Risk aversion avoids visible embarrassment. Therapeutic framing justifies softening feedback. Healthy embarrassment gets removed as a calibration tool, and no single actor decided to kill it.

Without calibration, orientation detaches from reality. Feedback loops collapse. The map is treated as the terrain because no mechanism forces comparison between the two. Initiative becomes dangerous because the system cannot absorb and correct the errors that initiative inevitably produces.

Mission Command requires a system that can see itself clearly and correct course in motion. Zero-defect culture, legal risk aversion, and the pathologizing of discomfort systematically dismantle that capacity.

Emotional Regulation Under Stress

In one civilian resistance network operating under sustained pressure, members were intentionally given wide latitude to act independently while at a public-facing action.[9] The environment was one where both uniformed and undercover law enforcement presence was assumed, and the potential for confrontation existed.

Several members carried an unacknowledged fear of exposure, arrest, social consequences, or failure. Others carried equally powerful validation needs: the desire to be seen as brave, committed, or indispensable. Both emotional states existed before decentralization occurred. Distributing authority allowed them to shape action without any constraint.

Members driven primarily by fear became risk-avoidant to the point of paralysis, choosing in the moment to not take part. The initiative collapsed because fear dominated judgment. Many of the same people who had been openly excited about the opportunity for action were seen functionally disengaging. One person had a panic attack at the thought of being arrested, even though they had not done anything to make that a possibility.

Other people moved in the opposite direction. Validation-seeking members interpreted autonomy as permission to escalate publicly and dramatically, or to refuse their assigned duties because they would be behind the scenes. Visibility became more important than effectiveness. Actions were chosen for their emotional payoff, attention, and affirmation, and a sense of purpose, rather than for their strategic value. These people took risks that compromised the operating environment for everyone else, drawing scrutiny and accelerating consequences.

In both cases, the core issue was the same: the absence of emotional regulation as a structural condition.

Under stress, unregulated fear narrows perception and overweights worst-case outcomes. Unregulated validation needs overweight recognition and underweight consequence. When either state drives decision-making, initiative becomes reactive rather than adaptive.

The distortions were invisible to the actors themselves. Fearful members believed they were being prudent. Validation-driven members believed they were being courageous. Both could justify their behavior using the language of commitment, caution, or moral urgency. Without a shared orientation and a structural recalibration mechanism, intent was interpreted through emotional need rather than operational reality.

Members who raised concerns about escalation were accused of cowardice or disloyalty. Members who questioned paralysis were accused of being provocateurs or undercover agents themselves. Because emotional states were never named or constrained structurally, disagreement became moralized. Once that happened, correction was no longer possible.

This is why emotional regulation is a prerequisite for Mission Command rather than a byproduct of it. Decentralized systems assume that individuals can hold fear, urgency, and uncertainty without allowing those emotions to hijack judgment. They assume actors can tolerate ambiguity without freezing and responsibility without seeking external validation. When those assumptions are false, autonomy becomes a force multiplier for emotional distortion.

This network distributed emotional instability along with the authority. The result was predictable: fragmentation, exposure, burnout, and collapse. Almost none of the groups taking part in that action still exist, and most of the members have severely curtailed their activism. Those who possessed the necessary self-regulation went underground.

Reliability Without Oversight

A volunteer organization in a rural region seemed to function well to the casual observer.[10] Members appeared dedicated. Public events were consistently well-staffed. The community supported the group financially and spoke highly of it.

Behind the public-facing operation was a different story. The real work of the organization, including training, maintenance, preparation, and quiet risk management, occurred almost entirely outside public view. It was there, away from donors and community praise, that reliability mattered.

Some members were genuinely dedicated, and found themselves carrying the bulk of the work. Others showed up exclusively for public-facing events and ignored the unglamorous tasks. Competence was uneven, but charisma in public obscured the gap. The system had been optimized for appearance rather than function, and over time, it drove out reliable members because true reliability imposed friction on a culture designed around visibility.

Mission Command assumes that when authority is delegated, actors will continue to meet standards even when no one is present to enforce them. In this organization, competence existed only where attention did. The moment visibility disappeared, so did discipline. Decentralizing authority in that environment would not have produced adaptability. It would have decentralized risk onto the members still trying to do the actual work.

Reliability without oversight is non-negotiable precisely because decentralized systems cannot supervise every decision. The doctrine assumes that standards are internalized rather than performed like an actor in a role. When they are merely performed, the system functions adequately under observation and degrades the moment observation ends. That degradation is invisible until it is not, and by then the conditions that produce Mission Command are destroyed.

Operational Closure

Mission Command requires five observable system properties. Shared orientation to reality, so that intent is interpreted against a common map. Internalized discipline, so that authority can be distributed without requiring external enforcement at every decision point. Healthy embarrassment and protected error correction, so that the feedback loop that allows a decentralized system to self-correct remains functional. Emotional regulation under stress, so that fear and validation needs do not drive initiative in the place of operational judgment. And reliability without oversight, so that competence survives when observation ends.

None of these conditions emerges naturally in most human systems. Each one is actively destroyed by mechanisms that modern institutions build and maintain for other reasons: career protection, legal risk management, cultural comfort, and the prioritization of appearance over function.

The question Part II now turns to is not whether these conditions are difficult to maintain. They are. But why are they so rarely present even in systems that believe they have built them? What does failure look like? How can it be identified?

What You Can Now Diagnose

- Whether your system shares a common operational map, or whether official narratives and lived realities have quietly diverged.
- Whether discipline in your system is internalized or externally enforced, and what happens to decision-making when external enforcement is unavailable or slow.
- Whether embarrassment produces recalibration in your system, or whether visible error triggers silence, self-protection, or informal punishment.
- Whether your after-action processes generate honest orientation or whether they generate polished accounts of what leadership prefers to have happened.
- Whether fear or validation needs are driving initiative in your system without being named or constrained.
- Whether competence in your organization survives when no one is paying attention.

Chapter Notes

1. The CIB patrol account is drawn from the firsthand testimony of E, an infantryman who provided his account anonymously to the authors.

2. NTC rotation observation drawn from Donald E. Vandergriff's training center work.

3. On Prussian post-battle critique culture and the Moltke General Staff system, see Robert M. Citino, *The German Way of War: From the Thirty Years' War to the Third Reich* (Lawrence: University Press of Kansas, 2005), 131--168; Bruce Gudmundsson, *Stormtroop Tactics: Innovation in the German Army, 1914--1918* (Westport, CT: Praeger, 1989), 3--18; Donald E. Vandergriff, Adopting Mission Command (Annapolis, MD: Naval Institute Press, 2019), 45--72.

4. Cory Zillig, "Anonymity is like Alcohol," Bellum Aeternum (Substack), August 2, 2024. https://substack.com/home/post/p-147292500

5. The pilot instrument analogy for healthy embarrassment appears in Donald E. Vandergriff's research material and is used here with his attribution.

6. On zero-defect culture and its effects on leader development and feedback, see Leonard Wong and Stephen J. Gerras, *Lying to Ourselves: Dishonesty in the Army Profession* (Carlisle, PA: U.S. Army War College Strategic Studies Institute, 2015); Donald E. Vandergriff, "The Tyranny of Metrics in Leader Development," Donald Vandergriff (Substack), March 2025.

7. On therapeutic reframing and the pathologizing of constructive discomfort in institutional settings, see "Trauma-Informed Leaders and the Behavioral Health Golden Hour," NCO Journal, October 31, 2023.

8. The resistance network case draws on Kit Perez's operational observation, 2014-2018.

9. The volunteer organization case draws on Kit Perez's observation, 2017-2024.

CHAPTER 5: THE PATTERN IS NOT UNIQUE TO THE MILITARY

The failures examined in Part I and Chapter 4 might sound like purely military problems, but they are structural problems that happen to have been documented most precisely in military contexts, because military systems face consequences that force documentation.

The same mechanisms appear in corporations, churches, activist organizations, and volunteer groups, producing the same degradation in the same sequence, for the same reasons.

What changes across domains is the vocabulary used to describe the system and the consequences when it fails. What does not change is the underlying architecture. Moral language replacing structure, flat hierarchies generating unaccountable power, and decentralization collapsing when orientation was never stable in the first place: these are predictable outcomes of the same structural absence.

The three sections that follow demonstrate this using composite cases drawn from observed patterns in civilian organizational environments.

Moral Language as Structural Substitute

In a mid-sized technology services firm, senior leadership undertook what they described as a "cultural transformation." The organization had grown quickly, promoted managers without leadership development, and accumulated a layer of middle management that coordinated activity without exercising real authority. A consulting engagement produced a new set of organizational values: ownership, initiative, transparency, and trust. These were posted on walls, incorporated into performance review language, and repeated in all-hands meetings.

The stated intent was to push decision-making downward and create a culture where everyone took responsibility for outcomes. In practice, the values language performed a different function. It signaled what kind of person belonged in the organization without specifying what that person was authorized to do or accountable for.

"Ownership" was invoked frequently and never defined clearly. Initiative was praised in the abstract and punished selectively in practice, depending on whether the outcome aligned with what leadership had privately preferred. Transparency was expected of subordinates and optional for leadership.

The language replaced structure because that was easier than building it. Defining authority, establishing clear constraints, specifying who decides what under which conditions, and creating protected mechanisms for surfacing error all require sustained effort and political will. Adopting values language requires nothing more than a workshop and a wall display.

The operational consequences were catastrophic. Three teams had each made decisions that contributed to a major product failure, each operating under the assumption that ownership meant they were empowered to act in their domain. None had a mechanism for surfacing the dependencies between their decisions. None had defined constraints on what they could authorize unilaterally.

When the failure emerged, each team's behavior was technically consistent with the values language. There was ownership and initiative, but transparency had not been practiced because no one had defined what transparency required structurally, only that it was a value the organization held.

The after-action process focused on whether individuals had demonstrated the values sufficiently rather than on whether the authority structure had been adequate for the decisions that needed to be made. Several individuals were counseled for "insufficient ownership." The structural gaps that had made the failure predictable were not examined, and leadership dodged the responsibility that was ultimately theirs.

This is what moral language as structural substitute looks like in operation. Values language signals virtue without requiring behavior. It provides a shared vocabulary that functions as social currency inside the organization, marking who belongs and who does not, without specifying what belonging requires in practice.

Misaligned actors can inhabit the language indefinitely because the language has no behavioral floor. Competent actors cannot rely on it because it provides no decision architecture. And when the inevitable failure happens, the language becomes the diagnostic tool, which means the diagnosis will always land on individuals rather than structure.

Mission Command in this environment was not possible. The prerequisites had been replaced by their linguistic shadows. The system could not self-correct because it had no mechanism for distinguishing between a person who embodied the values and a person who performed them. In the absence of that distinction, the performers thrived and the practitioners accumulated friction.

Flat Structures and Informal Power

A regional activist organization operated without formal leadership roles as a deliberate design choice. The founding members had watched other groups fragment under personality-driven leadership and had concluded that hierarchy was the problem. The solution, as they understood it, was to eliminate hierarchy and distribute authority equally across all the people in the group. Decisions would be made by consensus. No one would hold a title; the group would be held together by shared commitment to the mission.

Within six months, the group had a leader anyway. She held no title and had never been elected to any position. She also controlled which proposals reached the full group for discussion, set the emotional temperature of every meeting, and had made clear through several early incidents that members who challenged her preferred direction would face social consequences that ranged from public criticism to quiet exclusion from informal communications.

The mechanism was not complicated. When formal authority is removed, the functions that authority performs do not disappear. Decisions still need to be made, proposals still need to be filtered, and the group still needs someone to resolve disputes and set direction. In the absence of formal structure, those functions migrate to whoever is most willing to exercise them.

In this group, that person combined high emotional intensity (which made her positions feel like moral stakes rather than preferences), with a consistent willingness to apply social pressure to dissenters (which made disagreement feel costly). Both mechanisms reinforced each other. Her intensity signaled that the mission was at risk if her direction was not followed. Her social pressure demonstrated that the risk of opposing her was personal rather than merely political.

Members who disagreed with her direction faced a consistent double bind. Raising concerns openly was framed as divisive or insufficiently committed to the mission. Raising concerns privately produced the same social pressure in a smaller audience. Saying nothing was the path of least resistance, and most members took it.

Over time, the group's stated consensus masked a functional autocracy that was more insulated from accountability than any formal hierarchy would have been, because no mechanism existed to remove her, no authority structure could override her, and the group's foundational commitment to leaderlessness made naming the dynamic nearly impossible without appearing to contradict the organization's own values.

The orientation consequences were predictable. Shared orientation to reality requires that actors can surface uncomfortable truths without paying a personal cost. In this group, the only truths that could be surfaced safely were those that aligned with the de facto leader's existing position. Disagreement had been structurally converted into disloyalty. The group's map of reality was therefore the map that one person found acceptable, and that map was never tested against any corrective mechanism because no corrective mechanism existed.

Removing formal hierarchy produced unaccountable power held by the person most willing to exercise it without constraint. That power was more durable, more insulated, and more damaging to shared orientation than any formal authority structure would have been, because formal authority can be checked, appealed, and removed. Informal authority that exists inside an organization committed to its own non-existence cannot be named, let alone corrected.

This is why flat structures or "leaderless" organizations cannot produce the conditions Mission Command requires. They eliminate the accountability mechanisms that make authority correctable. What fills the vacuum is determined by emotional force and social willingness to punish dissent. Those are precisely the wrong selection criteria for a system that requires calibration to reality under pressure.

Decentralization and Internal Collapse

A software development firm operating in a competitive market made a deliberate strategic decision to adopt a decentralized organizational model. The company had grown to approximately two hundred employees and was struggling with the latency of centralized decision-making. Product teams waited weeks for approvals that should have taken hours. Engineering decisions escalated to leadership, who lacked the technical context to evaluate them.

The solution, as leadership framed it, was radical autonomy. Teams would own their domains entirely. Leadership would set direction and get out of the way.

The announcement was made at a company all-hands meeting. Teams were told they had full authority over their roadmaps, their technical decisions, and their resource allocation within broad parameters. Several teams responded immediately with energy and initiative. The first quarter after the announcement produced visible momentum.

What leadership had not examined, however, was whether the organization had the prerequisites that autonomy required. Teams had been operating under centralized control for years and had developed habits calibrated to that environment. Decisions had been deferred upward so consistently that many team leads had lost confidence in their own judgment for consequential choices.

Shared orientation across teams was shallow: each team understood its own domain but had limited visibility into how its decisions affected adjacent teams or the company's larger operational picture.

Internalized discipline was uneven. Some teams had strong norms around quality and constraint. Others had learned to optimize for what leadership had historically rewarded, which was speed and visible output rather than sound judgment.

Under centralized control, these gaps had been managed by the oversight structure itself. Approvals had caught some bad decisions before they propagated. Coordination requirements had forced some cross-team communication that would not otherwise have occurred. When that structure was removed, the gaps it had been managing became load-bearing.

Within a year, multiple significant failures had occurred. One team made a technical architecture decision that created dependencies that would cost eighteen months to untangle. A second team allocated resources to a feature set that directly contradicted the direction two other teams were moving, a conflict that would have been visible with minimal cross-team communication but that no one had surfaced. A third team, uncertain how to exercise its new authority, spent six weeks in internal deliberation over a decision that should have taken a day, missing a market window in the process.

Leadership's response was to reassert control. The autonomy initiative was quietly retired, described in subsequent communications as a learning experience. Several team leads who had struggled most visibly were counseled or replaced.

The diagnosis that circulated inside the company was that the teams had not been ready for autonomy. That diagnosis was correct in a narrow sense but missed the structural explanation entirely.

The teams had not been ready because the organization had never built the conditions that autonomy requires. Shared orientation had not been established before authority was distributed. Discipline had not been internalized because the prior system had never required it. Error correction mechanisms had not been created because the announcement of autonomy had been treated as the work rather than the beginning of the work.

Decentralization had been declared with assumption that the prerequisites were present. The system collapsed along exactly the lines where the prerequisites had been absent, and the collapse was blamed on the people rather than the structure that had never been built.

This is the same failure that appears in the EMS organization in Chapter 1, the resistance network in Chapter 3, and the advisory teams in Afghanistan. Domains, vocabulary, and organizational charts change, but the failure mechanism is the same.

When orientation is unstable, decentralization distributes risk. The faster the system moves under autonomy, the faster the misalignment propagates. The collapse, when it arrives, looks like a people problem because the people are where the consequences land. The structural absence that produced it remains invisible until someone is willing to look past the individuals to the conditions that were never built.

Operational Closure

Moral language performs alignment without producing it, which means misaligned actors can inhabit it without detection and aligned actors cannot rely on it under pressure.

Flat structures eliminate the accountability mechanisms that make power correctable, and the vacuum fills with whoever is most willing to exercise force without constraint.

Decentralization announced before orientation is stable distributes whatever misalignment the system was already carrying, faster and further than centralized control would have allowed.

These patterns are the predictable consequences of the same structural absence: prerequisites assumed rather than built, orientation unstable before authority was distributed, and no mechanism in place to surface and correct the misalignment before it propagated into collapse.

Part II has now established what Mission Command requires, and how systems across every domain destroy those requirements. The chapters that follow examine what failure looks like when those requirements degrade.

What You Can Now Diagnose

- Whether your organization uses values language to perform the accountability functions that structure should perform, and what happens when that language is tested by a failure that requires structural correction rather than a values conversation.
- Whether removing formal hierarchy in your system eliminated power or merely made it unaccountable, and who currently holds authority that the org chart does not reflect.
- Whether decentralization in your system was announced or built, and whether the prerequisites were present before authority was distributed or assumed to arrive alongside it.

- Whether the failures your organization attributes to individuals would survive scrutiny as structural problems if the conditions that were absent before the failure were examined honestly.

CHAPTER 6: INTERNAL CAPTURE AND SYSTEM DEGRADATION

Decentralized systems fail primarily through internal distortion rather than external attack. The threat is inside the orientation loop.

The failures examined in Part II share a common feature: they were all visible in retrospect and invisible in the moment. Organizations operating under assumed Mission Command do not experience themselves as degraded, and if asked would describe themselves as functional, committed, and aligned.

The gap between that self-perception and operational reality is the product of a specific degradation sequence that assumed Mission Command creates the conditions for and that internal capture exploits.

Assumed Mission Command generates vulnerability because it distributes authority into a system where orientation is unstable and error correction is structurally suppressed. Over time, that combination produces a system that actively defends its own misalignment. The three stages of that defense are narrative drift, emotional leverage, and competence displacement. Each stage enables the next. Together they constitute internal capture: the condition in which a system's orientation has been colonized by the story it tells about itself, rather than the reality it operates within.

The volunteer organization introduced in Chapter 4 provides the through-line for what follows. That chapter established the baseline: a system optimized for appearance over function, where competence existed only where attention did and degradation was invisible until it was not. Chapter 6 begins where Chapter 4 left off, at the point where the degradation that had been invisible becomes structural.

Narrative Drift

The volunteer organization had always maintained a particular description of itself: dedicated service, community trust, irreplaceable expertise. That account was not entirely false when the organization was founded. A core of genuinely capable and committed members had built something real, and the founding description accurately reflected them.

As those members aged out or departed under friction from a system that rewarded visibility over function, the account did not change with them. It remained in place, repeated at public events, donor meetings, and community gatherings, while the operational reality beneath it quietly shifted. The organization that the description claimed to represent was no longer the organization that existed. The institutional narrative had become self-referential: evidence that the organization was what it claimed to be was drawn from the narrative itself, rather than from operational performance.

This is the first stage of narrative drift. The system adopts a story that flatters rather than describes, and then leadership actively protects that story because it has become load-bearing. It is the basis for donor or customer relationships, community standing, and the self-concept of the members who remain. Challenging it feels like an attack on the organization rather than a calibration of its map against reality.

The protection mechanism is not always deliberate. In this organization, most members believed the institutional account sincerely. They had joined because of it, had been socialized into it, and experienced its repetition as confirmation rather than performance. The few members who held accurate assessments of operational capability found that surfacing those assessments produced social friction where the consequences were too severe. The narrative was defended as identity, and accurate assessments were categorized (and punished) as disloyalty.

This is where narrative drift becomes structurally dangerous. A system whose self-description cannot be challenged has severed the feedback loop that allows orientation to be calibrated against reality. Shared orientation now means shared commitment to the flattering account rather than shared perception of actual conditions. Intent issued inside that system is interpreted against the narrative map, not the operational map. The two maps have diverged, and no mechanism exists to force their comparison.

In Mission Command terms, the orientation failure that Chapter 4 identified as a precondition problem has now become a structural feature. The system is missing shared orientation, but now it has gone further and is actively producing and defending misorientation. That distinction changes the nature of the intervention required. Missing conditions can be built. Actively defended misorientation requires that the defense mechanism be dismantled before building can begin.

Emotional Leverage

Once narrative drift has produced a defended institutional narrative that leadership actively protects, the system requires a mechanism to manage actors who threaten that account. In the volunteer organization we are following, that mechanism operated through two simultaneous channels: punishment of members who surfaced accurate information, and reward of members who performed alignment with the narrative.

The punishment channel was rarely explicit. Members who raised concerns about operational capability, training gaps, or the behavior of legacy members were not typically told outright to be quiet. They were instead subjected to a consistent set of social responses. Their concerns were attributed to bad motives, personal grievances, or insufficient understanding of how the organization worked. Their standing in informal networks was quietly reduced. They were assigned to tasks that removed them from visibility and influence. Over time the message was clear without ever being openly stated: accurate assessment of the organization's actual condition would be seen as a problem.

The reward channel operated in parallel. Members who publicly affirmed the organization's narrative, who spoke well of the legacy members and performed enthusiasm and commitment at public events, received recognition, inclusion in informal decision-making, and advancement into visible roles. Their performance of alignment was treated as evidence of alignment. The distinction between the two was never examined, because examining it would have required questioning the narrative the rewards were designed to reinforce.

This is not groupthink. Groupthink produces blind spots through passive conformity pressure. What this describes is an *active suppression system*.[1] Members who introduced accurate information were functionally threatening to anyone whose position, standing, or self-concept depended on the inaccurate account remaining unchallenged. The social response was proportionate to that threat: deliberate, sustained, and distributed across enough of the membership that no single actor could be identified as responsible for it.

Together these channels constituted emotional leverage: the use of social reward and punishment to regulate what information could safely circulate inside the system. The leverage was distributed across the membership and enforced through the normal social dynamics of a group that had organized its identity around a particular institutional account. Any member who threatened that account threatened the identity of every member who had invested in it. The social response was therefore automatic and collective rather than calculated and individual.

The operational consequence was the completion of the orientation failure that narrative drift had begun. Narrative drift had severed the feedback loop between the flattering self-description and operational reality. Emotional leverage ensured that actors with accurate orientation could not safely introduce that accuracy into the system's decision inputs. The system was now not merely misoriented but actively defended against reorientation. Every mechanism that Mission Command requires for self-correction had been converted into a mechanism for self-protection.

Actors with accurate orientation faced a difficult choice: perform alignment with the narrative, suppress their assessment and continue operating, or leave. The first option required abandoning the honest engagement that made them valuable. The second was sustainable only for a limited time before the cognitive cost became prohibitive. The third option, departure, was the one most compatible with maintaining integrity. And departure was what the system, without intending to, was systematically selecting for.

Competence Displacement

The departure of members with accurate orientation was not seen by the organization as a loss, even though it should have been a signal that something was deeply wrong. Instead, it was framed as resolution. Members who had been sources of friction, who had raised uncomfortable questions and resisted the narrative, were gone. The remaining membership was more cohesive, more aligned, and more comfortable. The organization felt healthier than it had in years.

What had occurred was competence displacement: the systematic replacement of members whose value derived from accurate orientation and genuine capability with members whose value derived from performance of alignment. The organization had selected against the properties that made those members capable and selected for the properties that made them comfortable to have around.

The displacement operated through both channels identified in the previous section. Self-selection removed members who found the environment intolerable; once they realized they could not force the group to calibrate to reality, they self-selected out.

Active displacement removed members who persisted despite finding it intolerable; eventually they were either frozen out or summarily kicked from the group.

The two channels reinforced each other. As accurate members departed, the ratio of performers to practitioners shifted, which made the environment more hostile to the next accurate member, which accelerated the next departure.

This is the self-sealing property of competence displacement. Each departure makes the next one more likely. Each reduction in the ratio of practitioners to performers makes accurate orientation rarer and more threatening to the system's dominant narrative. The degradation accelerates, because the mechanism that would interrupt it, the presence of members with accurate orientation and the standing to surface it, is the mechanism being systematically removed.

In the volunteer organization, this process had been running for long enough that it had become generational. Leadership roles passed to people who had been socialized entirely within the organization's flattering narrative. They had no operational baseline against which to measure the current state because they had never seen the organization function at its actual historical capability. The story they had inherited described a level of competence that no longer existed and could no longer be verified. They repeated it sincerely because they had no alternative reference point.

By this stage, the organization had achieved what the original note in Chapter 4 described as the condition where reliable members had been driven out and the system continued operating on its own momentum. The system was in equilibrium, stable around a degraded state, defended by the members who had been selected to inhabit it, and invisible in its degradation to everyone inside it.

Even though their reputation had degraded along with everything else and they were repeatedly recognized as "a mess," the group was completely oblivious. Their story told them who they were even though it stopped being true a long time ago.

This is internal capture, the end-stage state of degradation. The orientation loop has been colonized. The inputs the system uses to assess its own condition are the outputs of the narrative the system is defending. There is no external reference point being consulted. There is no corrective mechanism operating. The system is generating its own confirmation and calling it shared orientation.[3]

Decentralizing authority into this system would not produce Mission Command, and announcing its adoption would distribute the degraded orientation faster and further. Introducing more trust, more empowerment, or more autonomy would accelerate the propagation of misalignment rather than correcting it. The prerequisites of Mission Command have been replaced by their functional inverses: a narrative that defends misorientation, a social system that punishes correction, and a membership selected for performance of alignment rather than accuracy of assessment.

What GCP Addresses

Part II has now traced a complete arc:

1. Mission Command requires specific conditions.

2. Those conditions do not emerge naturally and are actively destroyed by mechanisms that most human systems build and maintain for other reasons.

3. When the conditions are absent and authority is distributed anyway, the system degrades along predictable lines.

4. When the degradation runs long enough without correction, internal capture replaces the orientation loop with a self-referential narrative that the system cannot examine from within.

The Grey Cell Protocols are the answer to that specific problem, not to leadership failures in general, not to cultural dysfunction in the abstract, and not to the challenge of managing difficult people. GCP addresses the orientation loop directly.

GCP I interrupts competence displacement before it becomes self-sealing, by designing environments where misaligned actors expose and exit rather than accumulate.

GCP II builds structural trust that survives the betrayal, error, and social pressure that emotional leverage depends on.

GCP III provides the psycholinguistic tools to detect narrative drift before it becomes defended, by reading language as trace evidence of orientation state.

GCP IV maps the human archetypes through which internal capture most commonly operates, converting pattern recognition from instinct into diagnosis.

GCP V maintains the narrative integrity that keeps the orientation loop's inputs connected to reality rather than to the system's flattering self-description.

Together, they make it possible to detect the signs of (and path to) internal capture early enough to interrupt, and they build systems durable enough to survive the interruption without experiencing correction as collapse.

Part III delivers the operational doctrine for each facet. The reader now understands precisely what GCP is designed to prevent, and why each facet addresses the specific mechanism it addresses.

What You Can Now Diagnose

- Whether your organization's story about itself can be challenged from within, and what happens to members who try.
- Whether the members currently in visible leadership roles hold that position because of demonstrated capability, or because of demonstrated alignment with the dominant narrative.
- Whether the members your organization has lost in the past two years were disproportionately the ones who

raised uncomfortable questions, and whether their
departure was experienced as resolution rather than loss.

- Whether the social dynamics in your system reward
  accurate assessment or punish it, and whether that
  reward structure is explicit or operates through the
  informal channels that no one is responsible for and
  everyone enforces.
- Whether your organization's orientation loop is currently
  taking inputs from operational reality or from its own
  prior outputs.
- Whether, if you distributed more authority in your
  system today, you would be accelerating adaptability or
  accelerating the propagation of a misalignment that has
  not yet been named.

Chapter Notes

1. For more on groupthink in leadership contexts, see Kit Perez,
"How a Bad Leader Can Ruin Your Cause with Groupthink," *The
Shepard Scale* (Substack), May 14, 2021,
https://shepardscale.substack.com/p/how-a-bad-leader-can-
ruin-your-cause

2. The volunteer organization case in this chapter extends the
composite case introduced in Chapter 4. The degradation arc
described here draws on observed patterns across civic, volunteer,
and organizational environments over multiple years of
observation. Identifying details have been omitted throughout.

3. Kit Perez, "When the System Can't Diagnose Itself,\" *The
Shepard Scale* (Substack), March 8, 2026,
https://open.substack.com/pub/shepardscale/p/when-the-
system-cant-diagnose-itself

PART III: THE GREY CELL PROTOCOLS AND SEQUENTIAL USE

GCP is the infrastructure layer that creates and maintains the conditions Mission Command requires. Its five facets operate as a system. Mission Command and GCP are employed sequentially within a continuous reorientation loop.

CHAPTER 7: SELF-DISQUALIFYING SYSTEMS (GCP I)

The most survivable systems are designed so that misaligned actors expose themselves and exit. The burden of vetting then shifts from leadership vigilance to environmental design.

Part II traced how assumed Mission Command creates the conditions for internal capture. The failure patterns it described share a common entry point: misaligned actors accumulate inside the system before their misalignment becomes visible, and by the time it does, they have acquired enough leverage to defend their position against correction. The sequence is predictable because the conditions that allow it are structural, not incidental.

GCP I addresses the entry point directly. A self-disqualifying system is one designed so that the environment surfaces misalignment before it becomes entrenched. The actors who cannot operate within the system's actual demands identify themselves through their behavior under those demands, and they exit or are removed before they accumulate leverage. The vetting burden does not rest on leadership's ability to correctly evaluate character through interviews, references, or observation but on the structural properties of the environment itself.

These are two statements of the same claim. An environment designed to reveal what actors are under conditions that matter is, by definition, one that does not depend on leadership vigilance to sustain its integrity. The durability of the system is structural rather than personal. Leadership can be wrong, distracted, or replaced. The structure, however, persists.

The Inversion of Traditional Vetting

Traditional vetting assesses what people claim to be. References describe character. Interviews surface stated values. Credentials document prior performance in other contexts. The information gathered is real, but the problem it cannot solve is also real:

performance under conditions that do not yet exist cannot be assessed through processes that do not replicate those conditions.

A candidate who performs well in interviews has demonstrated the ability to perform well in interviews. A candidate with strong references has demonstrated the ability to sustain relationships with people willing to speak on their behalf. This is not useless information, but it also does not tell you how that person behaves when the environment is demanding, the stakes are real, authority is distributed, and no one is evaluating them.

Traditional vetting is also leadership-dependent, which makes it fragile in a specific way. The quality of vetting is a function of the quality of the leader doing it. A leader with poor judgment about character will vet poorly. A leader who is themselves misaligned will vet for alignment with their own orientation rather than the system's requirements. A leader who is overwhelmed, distracted, or operating under political pressure will vet inconsistently. The system's integrity at the entry point is therefore only as good as the leader currently responsible for it, which means it degrades every time that leader changes and must be rebuilt from scratch each time.

Self-disqualifying design inverts this. The environment itself carries the vetting function. A new leader inherits an environment with structural properties that surface misalignment regardless of whether the leader is attentive to that function or capable of performing it personally. The integrity of the entry point is embedded in the design rather than in any individual.

The Prussian General Staff system under Moltke the Elder operated on this principle, though it was not described in those terms.[1] Officers were not admitted to the General Staff through credential review alone. They were observed over extended periods under demanding conditions: war games, staff rides, exercises designed to surface how they reasoned under pressure, how they responded to being wrong, and whether their orientation to reality was stable enough to be trusted with consequential decisions. Rather than being a gate that candidates passed through, the vetting system was a sustained environment that revealed what they were. Candidates who could not sustain accurate orientation under those conditions identified themselves as unsuited for the roles the system needed filled. They did not need to be identified by an evaluator; the environment itself identified them.

In civilian terms, this is the difference between a probationary period designed as a formality and one designed as a genuine diagnostic. A probationary period that asks whether the new member shows up on time, follows procedures, and gets along with colleagues is assessing performance on low-stakes dimensions. A probationary period that introduces genuine pressure, requires decision-making under uncertainty, and creates conditions where misaligned actors must either perform their misalignment or suppress it at significant cognitive cost is assessing orientation. The first tells you whether someone can sustain compliance. The second tells you whether their map of reality is compatible with the system's actual demands.

Discipline as Identity: Culture as the Filter

The filtering function of a self-disqualifying environment depends on more than structural design. It depends on the culture that the design produces and sustains. Specifically, it depends on whether discipline is an enforced standard or an internalized identity, and on whether the culture makes that distinction legible.

When discipline is merely an enforced standard, actors comply because non-compliance produces consequences. The compliance is real, but it is contingent on the enforcement mechanism. Remove or weaken the enforcement, and compliance degrades. The actors who were complying for instrumental reasons find the calculation changed.

This is why systems that rely on oversight to produce discipline are fragile: they are disciplined only where and when oversight is present, which is precisely the condition that assumed Mission Command creates. Distributed authority reduces oversight by design. A system that was disciplined under oversight becomes undisciplined under autonomy, and the undiscipline was always latent.

When discipline is an internalized identity, the enforcement mechanism is secondary because actors do not experience compliance as what they do but as an expression of who they are. A unit that has genuinely internalized discipline does not require supervision to sustain it, because the members for whom discipline is identity will enforce it against each other without being directed to do so. The pressure is lateral and continuous rather than hierarchical and intermittent.

This is the filtering mechanism. An environment where discipline is genuinely held as identity is intolerable to actors for whom it is not. Those actors face a sustained choice between performing a discipline they do not hold, which is cognitively expensive and eventually unsustainable, or departing the environment where that performance is required. The environment simply needs to sustain the conditions that make performing misalignment more costly than exiting.

The 1-509th Infantry (Airborne), operating as the JRTC Opposing Force from 2018 to 2022, produced this dynamic deliberately.[2] Under commanders who treated disciplined initiative not as a performance standard but as the unit's operating identity, the unit consistently outperformed rotational forces that were larger and better resourced. The outperformance was a product of an environment where actors who could not sustain accurate orientation under pressure, who hedged, over-reported, or waited for direction when the situation required judgment, found themselves in friction with every lateral relationship in the unit. The culture enforced the standard without the command having to enforce it individually. Actors who could not sustain the identity either adapted or departed.

The civilian application is straightforward. Any organization that treats its operating standards as identity rather than compliance will produce lateral enforcement as a natural byproduct. The question for organizational design is how to make the standards legible and held deeply enough that the culture sustains them without hierarchical intervention. That is a design problem, not a management problem, and it has a different set of solutions.

Stress Testing as Orientation Diagnostic

Stress testing as a deliberate design feature and stress testing as an emergent byproduct of a demanding environment are distinct, and both are worth understanding because they produce different information and carry different risks.

Deliberate stress testing introduces conditions specifically to surface orientation before authority is distributed. The Reichswehr's interwar exercises were deliberate in this sense.[3] Officers were placed in scenarios designed not to train specific tactical skills but to reveal how they reasoned when their assumptions failed. The scenarios were constructed to make misalignment expensive: an officer whose orientation was calibrated to political appearance rather than operational reality would make different decisions than one whose orientation tracked actual conditions, and those decisions would be visible to evaluators and peers in a way that peacetime garrison behavior would not. The stress was the mechanism by which the training produced diagnostic information.

Emergent filtering is different. A genuinely demanding environment filters without being designed to filter. The demands of the environment are load-bearing in themselves, and actors whose orientation is incompatible with those demands accumulate friction that does not require external management. The filtering is a byproduct of the environment's actual requirements rather than a designed diagnostic feature.

Both are legitimate and they are often present simultaneously in the same system. The practical distinction matters because not all deliberate stress testing produces equally reliable diagnostic information. The variable that determines reliability is not duration but cost. A stress test that an actor can perform through by sustaining a surface presentation for its duration produces less reliable information than one whose demands exceed what most people can perform through artificially.

Basic Underwater Demolition/SEAL Training, or BUD/S, is time-limited at 24 weeks.[4] It is also among the most effective selection mechanisms in existence, precisely because the cost of performing orientation stability you do not hold is prohibitive within the course itself. Hell Week does not filter over months. It filters over days, through conditions where the gap between what an actor claims to be and what they are under extreme physical and psychological pressure becomes impossible to sustain through willpower alone. You cannot fake not quitting. The filter works because the cost of the performance exceeds the capacity of most actors to maintain it artificially, not because the course is long.

SERE (Survival, Evasion, Resistance, Escape) training operates on the same logic through a different mechanism.[5] The psychological pressure is calibrated to exceed the threshold at which most actors can perform composure rather than hold it. The course is short. The filter is reliable because the cost is extremely high.

The gameable version of deliberate stress testing is low-cost and observable: a structured interview, a one-day assessment, a single high-visibility exercise where actors know they are being evaluated and the stakes are limited. Those can be performed through, because the cost of maintaining a surface presentation for their duration is within the range of most motivated actors. The diagnostic information they produce reflects performance under assessment conditions, not orientation under operational ones.

The operational implication is that self-disqualifying design works best when deliberate diagnostic features are embedded inside environments whose demands are real rather than constructed. A selection course that exists solely as a filter is gameable by candidates who can sustain performance for the course's duration. A selection course that exists as the actual entry into the system's genuine demands, where the conditions do not change after selection is complete, is harder to game because the performance required does not end.

For civilian organizations, this means that stress testing as a standalone vetting mechanism is weaker than an environment whose operating culture genuinely sustains the demands that matter. The interview is weaker than the probationary period. The probationary period is weaker than a culture that sustains its standards after the formal probationary period ends. The strongest filtering mechanism is an environment that is simply what it claims to be, continuously, at every level of membership.

What a Self-Disqualifying Environment Looks Like in Operation

The positive case is straightforward to describe and consistently difficult to build. An environment is functioning as self-disqualifying when misaligned actors are departing or being removed before they accumulate leverage, when accurate information is circulating without significant suppression, and when the ratio of practitioners to performers is stable or increasing rather than degrading. None of these conditions require active management to maintain once they are structurally embedded. They are the natural output of an environment that is sustaining its demands honestly.

An anonymized armor company in 2024, drawing on the methods developed in Don Vandergriff's work on mission command restoration, produced this condition within a single year.[6] The company commander implemented sustained facilitated after-action reviews, protected members who surfaced accurate assessments from social and professional consequences, and treated honest acknowledgment of error as a professional standard rather than a personal weakness.

Within twelve months, the unit had moved from the bottom third of its battalion to the top third. Members attributed the change to an environment where failure could be examined without career cost, which meant failure was examined rather than concealed, which meant the unit's orientation to its own performance was accurate, which meant correction was possible. The self-disqualifying property operated through the AAR (after action review) culture: actors who could not sustain honest self-assessment in that environment identified themselves through their resistance to the process and were either developed or removed.

The failure of implementation is equally instructive and more common. Organizations that adopt the language of self-disqualifying design without the structural commitment produce a specific and recognizable failure mode. They announce high standards, describe themselves as demanding environments, and create formal processes that signal rigor. What they do not do is sustain the actual consequences that make the design function. Actors who fail to meet the standards are retained because removing them is costly, uncomfortable, or politically complicated. The standards become performative rather than load-bearing. The culture learns that the announced demands are negotiable, which means actors stop taking them seriously as filters, which means the misaligned actors the design was supposed to surface remain inside the system.

This failure mode is the organizational equivalent of the moral language pattern described in Chapter 5. The system adopts the description of a self-disqualifying environment without building one. The description signals virtue and sophistication. The structural commitment that would make it real is never made. The result is an environment that is more dangerous than one that never claimed the design at all, because the description provides cover for misaligned actors who can now point to the organization's stated standards as evidence of their own alignment.

The structural commitment that a self-disqualifying design requires is not complicated but it is non-negotiable: the consequences for misalignment must be real, consistent, and leadership-independent. Real means they actually occur, not that they are threatened. Consistent means they apply regardless of the actor's tenure, relationships, or current visibility. Leadership-independent means they persist when leadership changes, which requires that they be embedded in the culture and the process rather than in any individual leader's willingness to enforce them.

Where that commitment is present, the environment filters. Where it is absent, the environment is performing filtration while accumulating exactly the actors it was designed to remove.

Operational Closure

A self-disqualifying system as outlined in GCP I addresses the entry point of the degradation arc described in Chapter 6. [7] Narrative drift, emotional leverage, and competence displacement all require that misaligned actors be present inside the system long enough to accumulate leverage. Self-disqualifying design reduces that window by shifting the detection mechanism from leadership observation to environmental pressure.

It does not eliminate the need for the remaining GCP facets. A system that filters at entry still requires structural trust to survive the betrayal and error that will occur despite the best filtering design. It still requires psycholinguistic and archetype pattern recognition to detect misalignment that the entry-point design does not catch. And it still requires counter-narrative discipline to keep the system's self-description aligned with its actual demands rather than drifting toward the flattering account that undermines the filter.

Self-disqualifying design is the foundation. The chapters that follow build the structure that sustains it under operational pressure.

What You Can Now Diagnose

- Whether your organization's vetting process reveals how actors behave under conditions that match your actual operating demands, or whether it assesses performance on dimensions that are easier to measure and less relevant to what your system requires.
- Whether your operating standards are enforced from above or held as identity and enforced laterally, and whether you can distinguish between members who are compliant and members who are disciplined.
- Whether your stress testing is deliberate and diagnostic, emergent and genuine, or performative and gameable.
- Whether misaligned actors in your system are departing before they accumulate leverage, or whether your retention patterns suggest the filtering mechanism is not functioning.
- Whether your organization has adopted the description of a self-disqualifying environment without making the structural commitments that make it real, and whether your announced standards have real, consistent, leadership-independent consequences for non-compliance.

Chapter Notes

1. The Prussian General Staff vetting and development system draws on Donald E. Vandergriff, *Adopting Mission Command: Developing Leaders for a Superior Command Culture* (Annapolis, MD: Naval Institute Press, 2019); Robert M. Citino, *The German Way of War: From the Thirty Years' War to the Third Reich* (Lawrence: University Press of Kansas, 2005); and Bruce I. Gudmundsson, *Stormtroop Tactics: Innovation in the German Army, 1914--1918* (New York: Praeger, 1989).

2. The 1-509th Infantry (Airborne) JRTC Opposing Force account is drawn from Donald E. Vandergriff's observations. See also the aggregated unit climate surveys and after-action analyses

referenced in Don Vandergriff, various Substack posts, 2018--2025.

3. Reichswehr interwar training culture draws on Vandergriff, *Adopting Mission Command;* Citino, *The German Way of War,* and James S. Corum, *The Roots of Blitzkrieg: Hans von Seeckt and German Military Reform* (Lawrence: University Press of Kansas, 1992).

4. "Basic Underwater Demolition/SEAL (BUD/S) Training," NavySEALs.com, https://navyseals.com/buds/.

5. SERE training is administered in several versions across the branches of military service, varying in scope and intensity based on the personnel being trained. All versions share the core design logic of psychological and physical pressure calibrated to exceed performance thresholds. For general program overview, see U.S. Department of Defense, Joint Personnel Recovery Agency, SERE training program documentation.

6. The anonymized armor company case is drawn from Donald E. Vandergriff's mentoring and advisory work. Identifying details have been omitted. See Donald E. Vandergriff, Adopting Mission Command (2019), and the five-action restoration framework developed therein.

7. Kit Perez, "The Seven Operational Standards of a Self-Disqualifying Culture," The Shepard Scale (Substack), February 15, 2026,https://shepardscale.substack.com/p/the-seven-operational-standards-of.

CHAPTER 8: STRUCTURAL TRUST AND DAMAGE CONTAINMENT

Survivable systems do not rely on personal bonds. They are designed so that betrayal, error, and failure are absorbed structurally rather than experienced as collapse.

GCP I reduces the window during which misaligned actors can accumulate leverage, by designing environments that surface misalignment early. GCP II addresses a different problem: what happens when the filtering fails, when a trusted actor turns, or when a person the system was built around departs.

Every system will face these events. The question is whether the system was designed to absorb them.

Structural trust is the design answer to that question. It is an architectural property of a system: the degree to which the system's integrity is independent of the reliability of any individual actor within it. A system with high structural trust can absorb betrayal, error, and departure without experiencing collapse because its critical functions are not load-bearing on any single person's continued loyalty or presence. A system with low structural trust is only as resilient as its most trusted member's continued reliability, which means it is fragile in exactly the way that emotional trust cannot protect against.

The Brotherhood Illusion: Why Emotional Trust Is a Liability

Emotional trust is real and it is valuable. Units that have operated together under pressure develop bonds that produce genuine cohesion. Organizations built around shared commitment generate loyalty that sustains performance through difficulty. None of this is the problem.

The problem is when emotional trust is treated as a structural property rather than a social one. When a system's resilience is assumed to derive from the depth of its interpersonal bonds, two fragility vectors open simultaneously.

The first is betrayal. Emotional trust assumes continued alignment between the trusted actor's orientation and the system's requirements. When that alignment breaks, when a member's needs, beliefs, or external pressures shift, the trust that made them valuable becomes the mechanism of damage. The deeper the trust, the greater the access. The greater the access, the larger the attack surface when the trusted actor turns. Systems built on emotional trust, therefore, produce a specific structural vulnerability: the people most capable of damaging them are the people they have most thoroughly trusted.[1]

A composite drawn from observed patterns in civic and advocacy organizations illustrates this clearly.[2] An organization built over several years around a core of highly committed founding members operated on the assumption that shared mission and personal loyalty were sufficient structural foundations. Decisions were made informally among people who trusted each other. Access to sensitive operational details, donor relationships, and internal communications was distributed based on relationship rather than role. The founding members knew each other well, and that knowledge was treated as adequate vetting.[3]

When one founding member's orientation shifted, driven by a combination of external pressure and unmet personal needs that the organization had not examined or managed, the damage was proportional to the access the trust had granted. Donor relationships were compromised, and internal communications were exposed. The informal decision-making structure had no mechanism for isolating the damage because it had no mechanism for limiting access in the first place. The trust that had built the organization was the architecture through which it was damaged.

The second fragility vector is departure. A system built around the loyalty and capability of specific individuals is dependent on those individuals' continued presence. When a central figure leaves, whether through burnout, conflict, or opportunity, the system loses not just a person but the informal architecture that person sustained. Decisions that flowed through their relationships must now flow through different channels. Knowledge they held informally is no longer accessible. The cohesion they generated through personal presence must be rebuilt rather than maintained.

This is a different mechanism than betrayal, but the same underlying structural weakness. In both cases, the system's integrity was person-dependent rather than architecture-dependent. The brotherhood illusion is the belief that sufficiently deep personal bonds can substitute for structural design. They cannot, because no bond is permanent and no loyalty is unconditional, and a system that requires both to function is fragile in ways that no amount of trust-building can address.

In military terms, the distinction maps directly onto the difference between a unit whose cohesion depends on a single commanding officer's presence and one whose culture and processes sustain performance through leadership transitions.[4] The former is a common and understandable product of strong leadership. It is also a structural fragility that combat exploits. Remove the leader and the unit's performance degrades. A unit with structural trust sustains performance through leadership transitions because the standards, processes, and lateral relationships that produce performance are embedded in the culture rather than concentrated in any individual.

Structural Trust Defined: Compartmentalization, Redundancy, Process Over Personality

Structural trust is produced by three interlocking mechanisms that together make a system's integrity independent of any individual actor's continued reliability.

Compartmentalization limits what any actor can access and therefore limits what any actor can damage. The principle is straightforward: an actor who can only access what their current role requires can only compromise what their current role touches. This is based on the recognition that the actor's future reliability cannot be guaranteed, and that designing access accordingly protects both the system and the actor. A member who has never had access to group financials cannot expose them if their orientation shifts. A member who has never held communications infrastructure credentials cannot use them against the system if they depart under conflict.

Compartmentalization is standard practice in intelligence and special operations environments for this reason.[5] The principle is that operational security requires assuming the possibility of compromise and designing access structures that limit the damage any single compromise can produce. The same logic applies in any system where the cost of a single actor's betrayal or capture could be catastrophic.

Redundancy ensures that no single actor's failure is load-bearing. Critical functions, relationships, and knowledge must have backups that can operate independently. A donor relationship held exclusively by one member is a single point of failure. A process that only one person understands cannot survive that person's departure. A decision authority concentrated in one individual produces a system that cannot function when that individual is unavailable, compromised, or gone.

Redundancy is the cost of resilience. Systems that eliminate redundancy in the name of efficiency or 'tribal knowledge' discover its value when the single point of failure fails. The redundant capability that seemed wasteful or important becomes the only capability that remains.

Process over personality means that decisions are governed by structure rather than relationship. Who decides what, under which conditions, with what authority, and subject to what review: these questions have answers that do not depend on who currently holds a role or what relationships they have built. When a new person steps into a role, the process functions because the process is documented, enforced, and independent of the prior holder's personality and relationships.

This is the mechanism that makes structural trust durable across personnel change. Emotional trust must be rebuilt with each new actor. Structural trust persists because it is embedded in the design rather than in any individual. A system that governs decisions by process can absorb the departure of its most trusted members without losing its integrity, because the integrity was never located in those members to begin with.

Designing for Betrayal Before It Happens

The diagnostic question that anchors structural trust design is direct: if this member left or decided to destroy us tomorrow, how much damage could they do?

The question is uncomfortable because it requires treating trusted members as potential threat vectors. That discomfort is the point; a system that cannot ask this question about any of its members has allowed emotional trust to override structural analysis. The question allows an assessment of the system's current exposure. If the answer is "significant damage," the problem is not the member but the design.

Running this diagnostic against every member with meaningful access produces a damage exposure map. That map identifies where compartmentalization is insufficient, where single points of failure exist, and where process has been replaced by personality. It is the structural equivalent of a security audit, and it should be conducted with the same regularity and the same absence of sentimentality.

A composite case from an organizational consulting context illustrates what this looks like in practice.[6] A nonprofit operating in a politically sensitive environment had built its operations around three senior members who between them held all external relationships, all financial access, and all operational knowledge. The organization's leadership recognized the exposure after one of the three members left under conflict and the damage to donor relationships took eighteen months to repair. They commissioned a damage exposure audit before the remaining two members' situations were examined.

The audit identified that a single remaining member held exclusive relationships with the organization's five largest donors, exclusive access to the communications platform, and informal authority over hiring that had never been documented. The audit did not result in that member being treated with suspicion or even punished, because they had done nothing wrong. It did surface a structural problem within the group.

The audit--and the resulting conversations--meant that donor relationships were formally redistributed across multiple roles, communications access being restructured with documented protocols, and hiring authority being codified in a process that required two-person review. When that member departed eighteen months later under positive circumstances, the transition was absorbed without operational disruption. The structural design had converted a previously catastrophic exposure into a manageable personnel change.

The damage limitation principle applies identically in military contexts. Special operations units operating in denied environments design for the possibility that any member could be captured, compromised, or killed.[7] Access to mission-critical information is structured so that no single capture produces catastrophic exposure. That design is an acknowledgment that structural integrity cannot depend on the assumption that no member will ever be in a position where their loyalty is tested beyond what it can sustain.

In a structure based on likability or emotional trust, these types of conversations are seen as judgment or suspicion that might be emotionally harmful, and therefore avoided, to the group's potentially fatal detriment.

Damage Limitation as Routine System Architecture

Damage limitation has two modes, and both are necessary. Proactive damage limitation is the design work described above: compartmentalization, redundancy, and process structures that limit exposure before any failure occurs. Reactive damage limitation is the protocol that activates when failure has already occurred and the system must contain the damage that is already in motion.

Most organizations have neither. Systems in this position see betrayal, departure, or error as collapse rather than as a manageable event. The system has no pre-designed response, no clear authority structure for the crisis, and no documented process for isolating and containing the damage. The response is improvised under pressure, which is precisely the condition under which improvised responses produce the most additional damage.

Proactive damage limitation is architectural. It requires asking the damage exposure questions before they are urgent, building the compartmentalization and redundancy structures before any member's reliability has been questioned, and documenting the processes before any individual who currently holds them informally has departed. This work feels unnecessary when the system is functioning well, which is exactly when it must be done. A structure built during a crisis is a structure built under the worst possible conditions for clear thinking and careful design.

Reactive damage limitation is procedural. It requires having documented answers to the questions that a crisis generates: who has authority to isolate a compromised member's access, what is the protocol for notifying affected parties, who holds decision authority when the person who normally holds it is the source of the problem, and what is the recovery sequence for each critical function that has been disrupted. These answers cannot be developed in the moment because the moment does not allow for the deliberation they require.

The distinction between proactive and reactive damage limitation maps onto the distinction between resilience and recovery. Proactive design produces resilience: the capacity to absorb disruption without system failure. Reactive protocols produce recovery: the capacity to restore function after disruption has occurred. Both are required because proactive design cannot prevent all failures. A system that has both can absorb betrayal, error, and departure as operational events rather than existential crises.

In practice, the most common failure mode is organizations that have neither and discover the cost of that absence at the worst possible time. The second most common failure mode is organizations that have proactive design but no reactive protocol, which means they have reduced their exposure but have no structured response when exposure occurs despite the reduction. The rarest and most survivable configuration is organizations that treat both as routine architectural requirements rather than responses to specific incidents.

Operational Closure

GCP I filters misaligned actors before they accumulate leverage. GCP II ensures that when the filter fails, as it will, the damage is contained. The relationship between the two facets is not sequential in a simple sense. They operate simultaneously and each makes the other more effective.

A self-disqualifying system without structural trust is a filter that produces cohesion among the members who pass through it, but whose cohesion is built on emotional bonds that remain vulnerable to the fragility vectors described in this chapter. The filter reduces the probability of betrayal. It does not eliminate it, and it does not limit the damage when it occurs.

A structurally trusted system without a self-disqualifying design is a damage-containment architecture operating inside an environment that continues to accumulate misaligned actors. The containment limits the damage any single actor can do. It does not address the orientation degradation that results from the accumulation of performers over practitioners.

Together they address both the probability of failure and the cost of failure when it occurs. GCP I works on the probability. GCP II works on the cost. A system that has both can absorb the full range of human unreliability without experiencing it as collapse. That is the operational condition that Mission Command requires: a system designed so that the inevitable unreliability of human actors does not compromise the system's integrity or its orientation to reality.

What You Can Now Diagnose

- Whether your system's resilience is currently located in the reliability of specific individuals or in the architecture of the system itself, and what would happen to your critical functions if your two most trusted members departed tomorrow.
- Whether you can answer the damage exposure question for every member with meaningful access: if this person turned today, how much damage could they do, and is that answer acceptable.
- Whether your critical relationships, knowledge, and decision authorities have redundancy, or whether single points of failure exist that your system has not examined because the people currently holding them are trusted.
- Whether your decisions are governed by documented process or by informal relationship, and whether a new person stepping into any senior role would find a functioning process or a set of relationships they must rebuild from scratch.
- Whether your system has a reactive damage limitation protocol, and whether the people who would need to execute it in a crisis have read it, understand it, and know who holds authority when the normal authority structure is itself the source of the problem.

Chapter Notes

1. Kit Perez, "The Grey Cell Protocols: Structural Trust vs. Emotional Trust," The Shepard Scale (Substack), September 21, 2025, https://shepardscale.substack.com/p/the-gray-cell-protocols-structural.

2. The civic and advocacy organization case is a composite drawn from observed patterns across multiple organizational environments. Identifying details have been altered or omitted throughout.

3. For more on how familiarity substitutes for structural vetting, see Kit Perez, "When Liking Replaces Trust," The Shepard Scale (Substack), May 6, 2025,

https://shepardscale.substack.com/p/when-liking-replaces-trust.

4. The distinction between person-dependent and architecture-dependent unit cohesion draws on Donald E. Vandergriff, *Adopting Mission Command: Developing Leaders for a Superior Command Culture* (Annapolis, MD: Naval Institute Press, 2019), and the author's observations across military and civilian leadership development contexts.

5. Compartmentalization as standard intelligence and special operations practice draw on publicly available doctrine. See U.S. Army Field Manual 3-37.2, Antiterrorism (2011), and related operational security doctrine.

6. The nonprofit damage exposure audit case is a composite drawn from observed patterns in organizational consulting and advisory contexts. Identifying details have been altered or omitted throughout.

7. Special operations design for compromise and capture scenarios draws on publicly available doctrine and the authors' observations. Identifying details have been omitted throughout.

CHAPTER 9: PSYCHOLINGUISTICS, DECEPTION DETECTION, AND TRAUMA-INFORMED VETTING

Language is trace evidence of orientation. Trauma explains motive. Together, they give you an early warning system that operates before behavioral failure becomes visible.

GCP I designs environments that surface misaligned actors before they accumulate leverage. GCP II limits the damage when they are not surfaced in time. Both assume that misalignment is detectable. GCP III is the detection mechanism: the set of observational tools that reveal orientation drift before it produces behavioral failure visible enough to trigger the structural responses of the first two facets.

The detection problem is real. A skilled infiltrator can mimic discipline, and manipulators can 'perform loyalty' for months. Trauma-driven members can appear solid until unresolved needs drag the entire group off course. Traditional vetting checks ideology, field behavior, and stress responses. None of that reaches the level at which orientation operates: the level of emotional need.

Personal orientation gets warped internally, and that moves outward into the groups in which they involve themselves. Shame drives people to hide mistakes and lie about failures. Validation hunger drives oversharing that converts communications into a liability. People-pleasing produces compliance under manipulation rather than resistance to it. Anxiety amplifies rumors and accelerates rash decisions. Depression erodes reliability and forces the group to compensate for a member's absence. Each of these dynamics shifts a person's OODA loop from reality-tracking to need-management, and once that shift occurs, the group inherits the distortion.[1]

GCP III closes the detection gap through two interlocking tools. Psycholinguistic analysis converts language into observable orientation data. Trauma-informed vetting identifies the need structures that drive orientation drift before they are weaponized by an infiltrator or expressed as group-level damage. Neither tool requires clinical expertise, but both require disciplined attention.

One definitional clarification before proceeding. Infiltration, as used throughout this chapter and the doctrine it draws from, does not require an external actor with a deliberate agenda of disruption or even entrapment and arrest.

An infiltrator is anyone who is inside a group for a reason other than the stated mission. That includes the paid informant and the ideological plant, but it also includes the member who joined for social belonging, the member whose primary agenda is personal validation, the member who is managing an unresolved grievance, and the member whose needs have shifted since joining such that the mission is now secondary to those needs. The mechanism of damage is the same regardless of whether the misalignment was present at entry or developed over time. What matters operationally is the current orientation, not the original intent.

Why Psychology Is Non-Negotiable: Need-Based vs. Reality-Based Orientation

The distinction that anchors GCP III is operational rather than clinical. A person oriented to reality adapts under stress because their decision inputs track actual conditions. A person oriented to need processes the same conditions through the filter of what their unmet psychological needs require them to perceive. The external situation is identical. The map each person generates from it is critically different, and therefore the decisions arising from that map are different as well.

The operational consequences are predictable by need type:

- A person oriented to shame will deflect, lie, or hide rather than surface accurate information that might expose a failure.

- A person oriented to validation will perform and overshare rather than maintain the communication discipline the situation requires.
- A person oriented to approval will conform to manipulation rather than resist it, because the cost of resistance is social rejection, which their orientation cannot absorb.
- A person oriented to pity will exploit weakness to generate leverage rather than contribute to mission function.

These are operational liability assessments rather than character judgments. The intent of a shame-driven member may be entirely sincere. The effect is the same as deliberate sabotage: feedback loops break, errors get concealed, and the group's orientation to its own performance degrades. Intent does not change the operational consequence.

Every competent intelligence service recruits assets through this framework.[2] The FBI, for instance, does not need to manufacture weaknesses in a target organization; an infiltrator simply locates the member whose unmet needs make them a doorway and exploits those needs as the mechanism of access. The member who wants validation, fears rejection, or hides a secret past does not need to be turned. They need only be found and offered what they already want. The organization's psychological landscape is the attack surface. GCP III maps that surface before an external actor can use it.

The implication is direct: psychology and trauma-informed observation are not supplementary to doctrine but part of the core, because orientation is the system. If you cannot identify the inflection point at which a member has shifted from reality-based to need-based orientation, you will miss the moment at which your group begins to fracture. Infiltrators activate what is already present.

The Shame-Validation Scale as Operational Diagnostic

The primary diagnostic framework for need-based orientation is a sliding scale with shame at one end and validation need at the other. Both ends orient a person away from reality. Both produce predictable observable behaviors. Both are exploitable by anyone willing to meet the need the member is already expressing.

The Shame-Validation Scale

Shame End	Validation Need End
Hides mistakes	Overshares in communications
Refuses correction	Boasts about minor contributions
Silence when accountability is needed	Performs commitment for recognition
Withdrawal under stress	Exaggerate involvement or actions
Feeds cover stories and blind spots to infiltrators	Feeds information and credibility to infiltrators

Most people lean toward one end or the other, and under stress they slide harder in the direction they already lean. An infiltrator's task is simply to nudge them further in that direction. A shame-driven member will hand over cover stories and blind spots without being asked, because concealment is already their primary orientation. A validation-driven member will hand over information and credibility because being seen matters more than operational security.

The operational trap is that both ends of the scale present as loyalty. Shame-driven members look humble. Validation-driven members look committed and enthusiastic. Neither presentation is false exactly; it is real behavior generated by a need that will eventually override the mission requirements the behavior is currently serving. You can only catch the distinction if you are reading orientation rather than performance.

Mental health conditions that produce observable drift follow the same logic. They are not personal matters once they enter a group context. They become orientation signals, and they have predictable group-level costs.

Anxiety converts minor mismatches into crises. An anxiety-driven member becomes a rumor amplifier, escalating tension, generating urgency where none exists, and pushing rash decisions as a mechanism for relieving their own internal pressure. The group inherits their fear management instead of their reality tracking.

Depression produces disengagement. Deadlines slip, responsibilities go unmet, and communication drops off. The group compensates for the depressed member's absence, which redirects leadership energy from mission function to member management.

People-pleasing produces boundary collapse. A people-pleaser cannot say no to manipulation because the cost of refusal is the social rejection their orientation cannot absorb. They hand control to whoever pushes hardest, typically under the framing of keeping peace or maintaining group harmony. What's more, people-pleasers themselves engage in compulsive manipulation as a way to control what others think of them. This leads to dishonesty and other undesirable behaviors in the group.[3]

Shame produces secrecy. Once shame is in play, feedback loops inside the group cannot be trusted, because the shame-driven member will conceal exactly the failures the group needs to be able to examine and correct. Zero-defect cultures and those who select for careerism foster shame rather than healthy embarrassment.[4]

Validation addiction produces loose communications. The member who needs to be seen, praised, and acknowledged will eventually share details with outsiders to prove their worth, because the external recognition the group cannot provide is available from anyone willing to offer it.

None of this implies malice. In most cases, these members hold sincere commitment to the mission. The intent, however, does not matter if the effect is the same. The group that ignores these signals will treat liabilities as assets until the cost becomes visible, at which point the damage has already been done.

Psycholinguistic Markers: Pronoun Shifts, Tense Changes, Passive Voice, Linguistic Distancing

Language cannot be faked in the way that discipline, composure, and loyalty can. A skilled actor can sustain the performance of all three for months. What they cannot sustain is control over the subconscious orientation fingerprints their language leaves behind. Character, as trauma expert Diane Langberg has observed, forms in the interior before it manifests in visible behavior.[5] It surfaces in language first.

Every time a person speaks or writes under conditions that matter, they leave orientation markers: word choices and sentence structures that betray how they are mapping reality and what needs are driving that map. Most people do not recognize that their language is diagnostic. That is the operational advantage you will have.

The five primary markers are as follows.

Pronoun shifts occur when a speaker distances themselves from responsibility through grammatical construction. "That round ended up on the wrong steel target" attributes no agency to the shooter. "The police got tipped off" names no one who tipped them. These constructions tell you the speaker's orientation includes concealment of personal accountability. They will hide failures rather than own them.

Tense changes and time shifts occur when a speaker's temporal frame becomes unstable mid-account. "I was going to head over, and then I end up at a friend's place" shifts tense inside a single sentence, marking sensitivity around the transition being described. People often lose tense control at exactly the point in a narrative where fabrication or omission is occurring. Inconsistent tense does not prove deception; it simply marks the location where closer examination is warranted.

Irrelevant over-specifics occur when a speaker floods one portion of an account with unnecessary detail while leaving another portion vague or unexamined. Excessive detail about peripheral elements, what they were wearing, what vehicle they were driving, the incidental sequence of minor events, functions as a distraction structure. It fills space with verifiable truth while the significant gap in the account goes unexamined.

Passive voice and vague agents occur when a speaker removes themselves from causal chains. "The intel got leaked" names no leaker. "It just happened" assigns no actor. "Mistakes were made" is the grammatical form of maximum accountability evasion. Passive voice functions as a linguistic shield for shame, guilt, or deliberate deception. It is most diagnostic when it appears specifically around the events under examination.

Linguistic distancing occurs when a speaker avoids naming an event, object, or action directly, substituting vague reference instead. "That situation" instead of "the breach." "Someone made a call they shouldn't have" instead of "I told someone I shouldn't have." The distancing creates a gap between the speaker and the thing being discussed that their orientation requires maintaining.

Each marker in isolation is a signal that warrants attention, not a conclusion. Patterns across multiple markers in multiple exchanges are the diagnostic data. A speaker who consistently avoids personal pronouns, generates passive constructions around specific topics, and floods peripheral detail while leaving central questions vague is revealing an orientation map that does not align with the account they are presenting.

In Practice

The following list illustrates all five markers operating in sequence.[6]

- "Well, the meeting got pushed later than I thought." [Passive voice. Who pushed it?]
- "I was going to head over, and then I end up at a friend's place helping him move." [Tense change mid-sentence. What happened in the transition?]
- "I drove my truck, the silver Ford with the toolbox in back, the one I bought off Craigslist last year." [Irrelevant over-specific. The vehicle is not the question. The timeline is.]
- "Mistakes were made, but it's not like it mattered in the end." [Passive voice. The mistakes did not make themselves.]
- "That situation with the text thread, someone slipped, and it got around." [Linguistic distancing combined with passive voice. The speaker avoids naming themselves as the source of the leak, or is covering for whoever was.]

Every sentence in this account builds distance between the speaker and the facts under examination. Vague agents, over-specific peripherals, tense instability, passive constructions, and referential distancing appear in sequence across five consecutive responses. This is evidence of a pattern that reveals an orientation map that is organized around concealment rather than accuracy.

Whether the driver is shame, deliberate deception, or divided loyalty does not change the operational conclusion: this person cannot be trusted with consequential information or authority.

Trauma-Informed Profiles and the Seven-Question Needs Diagnostic

Language markers tell you that orientation has drifted. Trauma-informed profiles tell you why, which tells you how the drift will be exploited and what the group-level cost will be. The four primary profiles are not clinical categories and must not be used as such. They are operational patterns that produce recognizable behaviors, language signatures, and exploitation vectors.

The Four Trauma-Informed Profiles

The Approval-Seeker needs to be included in every decision and checks in constantly for confirmation that their standing is secure. The risk is collapse under manipulation: an infiltrator who offers praise, inclusion, and recognition converts the approval-seeker into a source of access without requiring any ideological conversion. The approval-seeker will not turn against the group overtly; they hand over access to whoever makes them feel most valued.

Language signature: "Am I doing this right?" "Do you still want me on this?" "Just wanted to make sure we're good."

The Validation Addict often overshares in group communications, repeats accounts of minor contributions, and brags about access or involvement. The risk is operational security: the validation addict leaks details to outsiders because external recognition satisfies a need the group's internal environment cannot fully meet. They do not intend to compromise the group. Being seen matters more than being secure, and that priority operates below the level of conscious decision.

Language signature: "I was the one who made that happen." "I already mentioned it to a few people so they'd know what we're about."

The Pity Gambler surfaces personal struggles to generate sympathy, reduce expectations, and obtain exceptions from group obligations. The risk is cohesion degradation: when pity becomes a transactional currency, standards stop applying to the wounded member, which creates a visible double standard that infiltrators and manipulators exploit. The group cannot maintain discipline when one member's suffering functions as a veto on accountability.

Language signature: "I wanted to be there; I've just been dealing with so much." "I know I said I'd handle it, but given everything..."

The Shame-Hider covers mistakes, becomes defensive when corrected, and deflects accountability through misdirection or counteraccusation. The risk is feedback loop collapse: a group that cannot surface and examine its own errors cannot adapt. The shame-hider ensures that specific failures remain unexamined, which means the conditions that produced those failures persist and recur.

Language signature: "That wasn't my fault." "It just happened that way." "Why is everyone focused on that?"

The Seven-Question Needs Diagnostic

The following questions are designed to surface need-driven orientation in both prospective members and existing ones. They are conversation prompts that generate the language samples the markers above are applied to. The questions themselves are less important than the language the candidate uses in response.[7]

1. Tell me about a time you made a significant mistake in a group or team context. What happened and what did you do?

2. What does it look like when you're under serious stress? How do people around you know?

3. Describe a time when someone in your group made a decision you disagreed with. How did you handle it?

4. What would make you feel like you were no longer valued in this group?

5. Tell me about a time you had to deliver bad news to someone whose opinion of you mattered. How did you approach it?

6. What's something you've failed at that you haven't fully resolved yet?

7. Describe a time when maintaining a boundary would cost you a relationship or standing with someone. What did you do?

Each question is designed to generate language around the need structures the four profiles produce. Question 1 surfaces shame orientation through the speaker's relationship to personal accountability. Question 4 surfaces approval and validation needs directly. Question 6 surfaces shame-hiding or validation-seeking depending on whether the speaker can name unresolved failure without deflection or performance. Question 7 surfaces people-pleasing and approval-seeking through the speaker's relationship to social cost.

Their answers are not necessarily the important segment; you are looking for answers in the language the answers generate. Apply the five markers above to each response. The patterns that emerge across the full set of seven questions are the orientation map you are building.

Operational Closure

GCP III is the early warning layer of the full protocol. GCP I and GCP II address what happens when misaligned actors are present inside the system. GCP III reduces the window during which misalignment goes undetected, by providing observational tools that operate before behavioral failure is visible enough to trigger structural responses.

The tools are sequential in practice:

- Psycholinguistic markers provide the first signal: language that reveals orientation drift before the drift has produced any visible behavioral failure.
- Trauma-informed profiles provide the explanatory layer: the need structure driving the drift, which predicts how the drift will escalate and how it will be exploited.
- The seven-question diagnostic provides the structured elicitation mechanism: the conversation format that generates the language samples the markers require.

Together they convert what most organizations try to manage through instinct, relationship, and retrospective analysis into a structured observational practice. The conversion matters because instinct is gameable, relationship is the attack surface, and retrospective analysis arrives after the damage has been done. GCP III moves the detection earlier, which is the only place where prevention is still possible.

Chapter 10 addresses what GCP III reveals once it has detected the pattern: the archetypes through which infiltration and internal capture most commonly operate, and the countermeasures each archetype requires.

What You Can Now Diagnose

- Whether you can identify, for each member with significant access or authority, where they sit on the shame-validation scale and which need structure is most likely to be exploited under pressure.
- Whether your vetting conversations are generating language samples rich enough to apply the five markers, or whether they are generating ideological declarations and behavioral self-reports that tell you what the candidate wants you to know, rather than what their orientation is.
- Whether you can identify the trauma-informed profile that best describes the members in your system who have been sources of friction, information leakage, or accountability resistance, and whether the pattern was visible before the damage occurred.
- Whether your current members' language under stress matches the orientation markers, and whether you have been reading those signals as personality quirks rather than operational data.
- Whether the need structures present in your group have been mapped, and whether you have identified which of them represents the highest exploitation risk for an external actor willing to offer what those needs require.

Chapter Notes

1. The foundational framework of need-based versus reality-based orientation and its group-level consequences is drawn from Kit Perez, "From Instinct to Evidence," The Shepard Scale (Substack), September 21, 2025, https://shepardscale.substack.com/p/from-instinct-to-evidence-trauma.

2. The FBI asset recruitment framework as an illustration of need-exploitation is drawn from Will Grigg's reporting on the William Keebler case. See Will Grigg, "Bill Keebler and the FBI's Entrapment Elves," The Libertarian Institute, June 27, 2016. https://libertarianinstitute.org/everything-will/bill-keebler-and-the-fbis-entrapment-elves/.

3. For more on people-pleasing as a form of compulsive manipulation and its effects on group cohesion, see Kit Perez, "The Narcissism of People-Pleasing: A Silent Threat to Your Group," *The Shepard Scale* (Substack), August 1, 2025, https://shepardscale.substack.com/p/the-narcissism-of-people-pleasing.

4. On zero-defect culture and its punitive effects on honest self-critique, see Leonard Wong and Stephen J. Gerras, *Lying to Ourselves: Dishonesty in the Army Profession* (Carlisle, PA: U.S. Army War College Strategic Studies Institute, 2015), 1--28, https://press.armywarcollege.edu/cgi/viewcontent.cgi?article=1465&context=monographs. On careerism as a driver of shame-based concealment rather than healthy embarrassment, see Donald E. Vandergriff, *Adopting Mission Command: Developing Leaders for a Superior Command Culture* (Annapolis, MD: Naval Institute Press, 2019), 45--72.

5. On deception detection and the SCAN method, see Avinoam Sapir, *The SCAN Technique* (Phoenix, AZ: Laboratory for Scientific Interrogation, 1987). For academic research on verbal deception cues including pronoun use and information density, see Aldert Vrij, Detecting Lies and Deceit: Pitfalls and Opportunities, 2nd ed. (Chichester, UK: Wiley, 2008).

6.   Diane Langberg, In Our Lives First: Meditations for Counselors (Jenkintown, PA: Diane Langberg, 2014). The observation that character forms in the interior before it manifests in visible behavior grounds the psycholinguistic premise that language surfaces orientation before behavior does.

7. The vetting conversation is adapted from Kit Perez, "From Instinct to Evidence," The Shepard Scale (Substack), https://shepardscale.substack.com/p/from-instinct-to-evidence-trauma.

8. The Seven-Question Needs Diagnostic is drawn from Kit Perez, \"From Instinct to Evidence." (Substack) https://shepardscale.substack.com/p/from-instinct-to-evidence-trauma.

CHAPTER 10: INFILTRATION ARCHETYPES

Infiltration is predictable. Naming the archetype shifts the group from gut instinct to pattern recognition and from suspicion to diagnosis.

GCP III provides the detection tools: psycholinguistic markers that surface orientation drift through language, and trauma-informed profiles that identify the need structures driving that drift. GCP IV provides the classification system: six archetypes that codify the roles destructive actors play inside groups, each with a recognizable behavioral pattern, a predictable operational aim, and a specific countermeasure that removes their leverage.

Proper classification is important, because suspicion without pattern recognition is operationally useless. A group that senses something is wrong but cannot name what it is will spend its energy managing anxiety rather than addressing the actual problem. Worse, undirected suspicion becomes a toxin that corrodes cohesion: members begin to distrust each other across the board rather than examining a specific actor's specific behaviors against a specific diagnostic framework.

The archetypes convert that diffuse suspicion into actionable diagnosis. The shift is from "something feels off about this person" to "this person is running a recognizable script, and we know how that script operates and how to neutralize it."

A definitional point carried forward from Chapter 9 applies here as well. An infiltrator is anyone inside a group for a reason other than the stated mission. The six archetypes are not primarily descriptions of government agents or paid informants, though any of them can be. They are descriptions of human psychological roles that appear in every organizational context: military units, corporations, civic organizations, activist groups, churches, and volunteer networks. The patterns are domain-independent because the psychology driving them does not change by context.

Each archetype is described below across ten diagnostic attributes: behavioral patterns, emotional needs, operational aim, grooming behavior, leverage points, communication signature, exploitation vector, lifecycle, collateral damage pattern, and detection latency. These attributes are woven into the prose description rather than presented as a checklist, because the diagnostic value comes from understanding how they interlock rather than from scanning a list.[1]

The Six Archetypes

1. The Pathological Ally

The Pathological Ally enters the group bearing gifts. Money, gear, weapons, training offers, logistical support: whatever the group currently lacks, the Pathological Ally arrives with it. The gifts generate rapid trust because they appear to demonstrate commitment. In reality they purchase access and establish a debt dynamic that will be activated later.

The operational aim is escalation. The Pathological Ally's orientation is organized around urgency: speed and reckless action are the only metrics of loyalty in their frame, and they will apply consistent pressure to accelerate the group's tempo beyond what its mission, planning, or security discipline supports. Restraint will be reframed as cowardice. Systematic decision-making is reframed as hesitation. Anyone who slows the pace is implicitly or explicitly cast as uncommitted or even cowardly.

The emotional need driving this is adrenaline and control. The Pathological Ally is not primarily ideologically motivated. They need escalating operational tempo the way a validation addict needs recognition: it is the need itself, not the mission, that is being served. Their grooming behavior reflects this: they build dependency quickly through gift-giving and enthusiasm, position themselves as the most capable actor in the room, and use that position to redirect the group's decision cycle toward their preferred tempo.

The communication signature is urgency framing combined with competence display. They speak in compressed timelines, frame every decision as time-critical, and consistently imply that the group's current pace reflects inadequacy rather than discipline. Under psycholinguistic examination, their language around restraint will reveal contempt: hedging language by others will be met with dismissal, and calls for systematic process will be characterized using distancing language that removes their legitimacy.

The exploitation vector is the group's desire to appear capable and committed. A group that measures its seriousness by operational tempo rather than orientation accuracy is a natural host for this archetype. The lifecycle moves from entry through rapid trust-building to role consolidation as informal operational lead, then to escalating pressure, then to either a precipitating event or exposure. Collateral damage typically includes legal exposure, blown cover, and reputational damage that survives the group's dissolution. Detection latency without a system in place is four to ten weeks.

One Pathological Ally joined a local property rights group that had been operating cautiously and effectively for two years.[2] Within six weeks of his arrival, the group was discussing direct confrontations that had never been part of their stated approach. Two members who raised objections were socially marginalized. The group's planning meetings shifted from structured agenda to reactive sessions organized around the Pathological Ally's latest proposed escalation. The group fragmented within four months, producing no outcomes and significant personal exposure for its members.

Countermeasure: Refuse urgency as a proof of loyalty. Discipline is the only valid currency in a mission-oriented group. Any actor who consistently frames systematic decision-making as cowardice or hesitation is revealing an orientation organized around something other than the mission's success. Force every significant decision through documented process regardless of the pressure to accelerate. The Pathological Ally has no leverage in a group whose identity is calibration rather than tempo.

2. The False Prophet

The False Prophet operates through narrative capture. Where the Pathological Ally uses urgency, the False Prophet uses ideology: charisma, rhetorical fluency, and apparent conviction deployed to gradually displace the group's original orientation with their own.

The operational aim is not usually the destruction of the group but the redirection of it. The False Prophet does not typically challenge doctrine directly. They add to it: new truths, refined definitions, reframed missions. The changes are incremental and each one is wrapped in the language of the group's existing values, which makes them difficult to challenge without appearing to challenge the values themselves. Over time the group's orientation has drifted substantially from its baseline, but because the drift was gradual and always expressed in familiar language, no single moment of departure is visible.

The emotional need is status and ideological primacy. The False Prophet needs to be the authoritative interpreter of the group's mission. Their grooming behavior is intellectual: they position themselves as the most sophisticated reader of the group's own doctrine, offer clarifying interpretations that subtly elevate their own authority, and reward members who adopt their framing with recognition and inclusion. Members who challenge the interpretations are characterized as missing the deeper point.

The communication signature is rhetorical sophistication combined with definitional drift. Under psycholinguistic examination, their language will show consistent pattern of redefining terms: words that the group uses in one sense will appear in the False Prophet's speech with slightly altered meaning, and the alteration will be presented as clarification rather than change. Tense and agency markers will be stable because the False

Prophet is not concealing behavior; they are constructing an alternative reality that they genuinely inhabit.

The exploitation vector is the group's ideological identity. A group that derives its cohesion from shared belief rather than shared mission is particularly vulnerable because the False Prophet can operate entirely within the belief structure while redirecting the mission. The lifecycle runs from entry through intellectual positioning to interpretive authority to narrative displacement. Collateral damage is typically mission drift, member polarization, and the departure of members whose orientation was most firmly anchored to the original baseline. Detection latency is long, often six months to a year, because the drift is gradual and each increment is deniable.

The Malheur National Wildlife Refuge standoff in 2016 illustrates this archetype at operational scale.[3] Participants who arrived at the initial rally understood themselves to be expressing support for the Hammond family's legal situation. The event was redirected through a series of narrative reframings, each expressed in the language of the group's existing values, until a significant portion of the participants had been moved to occupy a federal facility under conditions that served a different agenda than the one they had assembled to support. The redirection followed the False Prophet's script precisely.

Countermeasure: Structural clarity enforced through regular narrative audits. The group must have a documented baseline doctrine that new interpretations are explicitly tested against. Any shift in definition, framing, or mission scope requires explicit examination and group ratification rather than gradual acceptance through rhetorical accumulation. The False Prophet cannot bend a story that is anchored to a written, regularly revisited baseline. Challenge culture, discussed in Chapter 11, is the operating environment this countermeasure requires.

3. The Emotional Extortionist

The Emotional Extortionist leverages trauma and manufactured crisis to make their personal needs the group's primary operational concern. Their orientation is validation at any cost, and they require a continuous supply of it.

The operational aim is dependency creation. The Emotional Extortionist does not typically seek to expose or destroy the group. They seek to become indispensable to it through emotional leverage: guilt, pity, and manufactured urgency around their personal circumstances. Every group crisis passes through their frame. Every task is filtered through their need for attention or accommodation. Over time the group finds itself managing one person's emotional state as its primary activity.

The emotional need is pity and validation, often rooted in genuine unresolved trauma that has been converted into an operational instrument. The grooming behavior is vulnerability display: early disclosure of personal suffering, which generates sympathy and creates an implicit obligation in members who receive it. The Emotional Extortionist identifies which members are most susceptible to guilt and most uncomfortable with conflict, and those members become their primary leverage points within the group.

The communication signature is crisis language combined with boundary-testing. Their written and verbal communications will show high frequency of personal distress framing, implicit or explicit appeals to the group's obligation to accommodate them, and escalating emotional intensity when accommodation is delayed or refused. Under psycholinguistic examination, their language around the mission will show significant pronoun instability: the mission will drift into being framed around their participation in it rather than around its own objectives.

The exploitation vector is the group's discomfort with interpersonal conflict and its tendency to conflate emotional accommodation with cohesion. A group that mistakes pity for loyalty will reorganize itself around the Emotional Extortionist's needs without recognizing the reorganization as a structural compromise. The lifecycle runs from vulnerability display through sympathy generation to leverage consolidation to dependency loops that are progressively harder to break. Collateral damage is cohesion collapse, leadership energy depletion, and the departure of mission-oriented members who recognize that the group has been captured by one person's emotional management. Detection latency is typically eight to sixteen weeks.

An Emotional Extortionist joined a legal defense network that had been operating effectively for eighteen months.[4] Within two months, leadership meetings were regularly consumed by her personal crises. Members who raised operational concerns were made to feel that they were prioritizing the mission over a suffering colleague. Several high-performing members departed within four months, citing the group's inability to maintain mission focus. The network's operational capacity was reduced significantly before the pattern was formally named and addressed.

Countermeasure: Structural trust denies pity its leverage. A system in which roles are defined by function rather than personality, and in which no individual is indispensable, removes the mechanism the Emotional Extortionist requires. Pity cannot purchase operational exceptions in a system where exceptions are not available. When the leverage point is removed, the Emotional Extortionist either reorients toward the mission or self-selects out, which is the preferable outcome. Accommodation that preserves their presence while compromising the group's function is not a solution.

4. The Controlled Opposition Plant

The Controlled Opposition Plant is the most subtle archetype because their primary weapon is process rather than personality. Their operational aim is paralysis: the prevention of action through the endless management of alternatives, objections, and procedural concerns.

The behavioral pattern is consistent constructive obstruction. The Controlled Opposition Plant is always present, always engaged, always offering alternatives framed as improvements. They rarely oppose directly; they instead complicate things. Every plan acquires new risks that require further analysis. Every decision point generates new options that deserve consideration. Every meeting runs long because the process of evaluation never reaches conclusion. The group mistakes the appearance of thoroughness for actual deliberation.

The emotional need is control, though it may present as conscientiousness or caution. Their grooming behavior is helpfulness: they volunteer reliably, take on administrative roles, and position themselves as the voice of careful judgment. This positioning makes their obstruction difficult to challenge because it is consistently framed as responsible process management rather than opposition. The communication signature is hedge language combined with procedural expansion: "before we decide, shouldn't we also consider," "I just want to make sure we've thought it through," "what if we tried a lower-risk version first."

The exploitation vector is the group's legitimate valuation of deliberation. A group that has been correctly taught to avoid reckless escalation is susceptible to the Controlled Opposition Plant precisely because their behavior superficially resembles careful orientation. The distinction is that careful orientation produces decisions; the Controlled Opposition Plant produces only the appearance of careful orientation while preventing its outputs. The lifecycle runs from entry through reliability-building to procedural authority to systematic paralysis. Collateral damage is mission stagnation, member frustration, and the departure of action-oriented members who conclude the group is incapable of execution. Detection latency is long, typically three to six months, because each individual instance of obstruction is deniable as legitimate concern.

A Controlled Opposition Plant joined a coalition that had been building toward a specific regulatory intervention.[5] Over seven months, every proposed action generated a procedural objection that required resolution before the action could proceed. The coalition's meeting notes from that period show thirty-two scheduled decision points that produced no decisions. The intervention window closed. The coalition dissolved within a few months of the missed deadline, its members concluding that the group had been incapable rather than deliberately stalled.

Countermeasure: Bias for action built into the group's operating structure. Decisions made in advance through documented protocols do not require re-adjudication in the meeting where they would be applied. The Controlled Opposition Plant requires an open deliberation space in which to operate. Remove that space through advance decision architecture and the archetype has no mechanism. If an actor consistently generates procedural expansion regardless of the decision context, that pattern is itself diagnostic: legitimate caution produces decisions, and an actor whose caution never produces decisions is revealing their actual orientation.

5. The Drama Conduit

The Drama Conduit is frequently generated internally rather than inserted externally. Their orientation is chaos, and while they will consistently describe themselves as people who hate drama, they will consistently be found at the center of whatever conflict is currently consuming the group's attention.

The operational aim is environment generation: the Drama Conduit requires a highly activated emotional environment to feel oriented, and they produce that environment through gossip, rumor, and the amplification of interpersonal conflict. Every perceived slight becomes a grievance. Every disagreement becomes a fracture. Every piece of incomplete information becomes a story that travels through the group's side channels with additions and distortions at each transmission.

The emotional need is stimulation and informal power. The Drama Conduit accumulates influence through information brokerage: they know things others do not, and the currency of that knowledge is social standing within the group's informal hierarchy. Their grooming behavior is intimacy manufacture: they build close relationships quickly through shared confidences, position themselves as trusted confidants, and use those positions to generate and distribute the interpersonal friction their orientation requires.

The communication signature is high-frequency private messaging combined with loaded framing. Under psycholinguistic examination, their communications will show consistent use of implication over statement: "I probably shouldn't say this, but," "I don't know what's going on with so-and-so lately," "did you hear about..." The information they transmit is real information processed through a distortion lens that maximizes emotional activation. This makes it particularly difficult to address because the content is at least partially true and the distortion is in the framing rather than the facts.

The exploitation vector is the group's informal communication channels and its tolerance for interpersonal processing during operational discussions. A group that has no discipline around side-channel communication is a natural host for this archetype. Notably, external actors do not need to insert a Drama Conduit: most groups generate their own.

Governments and hostile organizations understand this; in many documented cases of group disruption, no infiltration was necessary because the group's internal Drama Conduits accomplished the disruption without assistance. The lifecycle runs from intimacy-building through information accumulation to conflict generation to group fracture. Collateral damage is cohesion destruction, leadership credibility damage, and the departure of members who require a functional operating environment. Detection latency is four to eight weeks in groups without communication discipline.

Countermeasure: Cultural clarity enforced through communication discipline. A self-disqualification environment in which gossip and rumor are treated as non-negotiable grounds for removal eliminates the Drama Conduit's operating space. The countermeasure must be structural rather than interpersonal: addressing a Drama Conduit through direct confrontation typically generates more material for the conflict architecture they require. The standard must be clear, documented, and applied without exception. A group whose culture rewards orientation to reality rather than social activation gives the Drama Conduit nothing to work with.

6. The Professional Plant

The Professional Plant is the archetype groups fear most and recognize least. Where the other five archetypes operate through personality, need, and emotional leverage, the Professional Plant operates through discipline. They are trained, patient, and oriented to mission rather than personal need. They do not require validation in the group. They show up consistently, perform competently, and wait.

The operational aim is structural compromise: leadership visibility, network mapping, access to plans and resources, and influence over outcomes sufficient to produce the desired result, which may be arrest, dissolution, or simply persistent operational failure. The Professional Plant is not necessarily seeking a dramatic outcome. Incremental damage that persists over time serves their mission equally well.

The grooming behavior is competence and reliability. The Professional Plant volunteers for positions that maximize their access to the information their actual mission requires. They are calm under pressure, which makes them appear trustworthy. They are disciplined, which makes them appear loyal. They are competent, which makes the group reluctant to scrutinize them because scrutiny feels ungrateful. Their communication signature is notably clean: under psycholinguistic examination, they will show few of the markers associated with deception because they are not concealing personal need. They are executing professional tradecraft, which produces different linguistic patterns than shame, validation hunger, or guilt.

The exploitation vector is the group's tendency to mistake competence and calm demeanor for proof of alignment. A group that grants access based on demonstrated capability rather than structural protocol is a natural host for this archetype. The lifecycle is the longest of the six: entry, slow trust-building over months, role consolidation into positions of maximum informational access, and then either a triggering event when sufficient intelligence has been gathered or indefinite persistence as a passive collection asset. Collateral damage includes arrests, network exposure, and the long-term chilling effect on member recruitment that follows a confirmed professional infiltration. Detection latency is the highest of the six archetypes, typically six months to several years without structural countermeasures in place.

In a well-documented case from the animal rights movement, a professional infiltrator maintained membership in multiple organizations over several years, attending meetings, building relationships, and accumulating organizational intelligence before his role was exposed.[7] His exposure came not through behavioral anomaly but through external documentary evidence. Within the groups he had infiltrated, his behavior had generated no significant suspicion because it matched the profile of an exceptionally reliable member rather than the stereotyped profile of a government agent.

Countermeasure: System, not suspicion. Compartmentalization ensures no single member sees the whole picture regardless of their tenure or demonstrated reliability. Impartial auditing applies security protocols to every member without exception: competence and charm cannot excuse anyone from structural norms. A culture of challenge normalizes questioning roles and access as a matter of standard practice rather than marking it as disloyalty. The Professional Plant can only operate in a group that mistakes discipline for proof of loyalty.

Structural trust, as defined in Chapter 8, denies them that advantage because the structural trust framework does not require proof of loyalty from any individual; it requires compliance with architecture that limits what any individual can access or damage regardless of their loyalty status.

How the Archetypes Interact with GCP I and GCP II

The six archetypes are variants of a single underlying problem: actors whose orientation is organized around something other than the group's stated mission, operating inside a group whose architecture does not systematically limit the damage that misaligned orientation can produce.

GCP I addresses probability. A self-disqualifying environment surfaces misalignment early and makes it expensive to sustain. The Pathological Ally finds no crowd willing to trade discipline for urgency. The Drama Conduit finds no informal social architecture to colonize because the culture does not reward the behaviors that architecture requires. The Emotional Extortionist finds that performing mission orientation under the consistent pressure of a demanding environment is a cost they cannot sustain. These archetypes are most effectively neutralized before they consolidate, and GCP I is the mechanism that prevents consolidation by making the environment hostile to their operating modes from the start.

GCP II addresses cost. A structurally trusted system limits what any actor can access or damage regardless of how long they have been present or how much trust they have accumulated. The False Prophet cannot redirect a narrative that is structurally anchored to documented baseline doctrine subject to regular audit. The Professional Plant cannot achieve structural compromise in a compartmentalized system where no single member's access includes the full operational picture. The Controlled Opposition Plant cannot paralyze a system whose decision architecture does not require their participation to produce outputs. These archetypes are most dangerous when they have accumulated leverage, and GCP II limits the leverage any single actor can accumulate by design.

GCP III provides the detection layer that operates between GCP I's environmental pressure and GCP II's architectural limits. Psycholinguistic markers surface the communication signatures each archetype produces. The Drama Conduit's implication-heavy private messaging, the Emotional Extortionist's crisis framing and pronoun instability around the mission, the Pathological Ally's urgency framing and contempt language around restraint: these are detectable patterns that GCP III's marker framework converts from vague unease into diagnostic data.

GCP IV names the patterns. Once named, the archetype's script loses much of its power because the group can see the role being played rather than experiencing its effects as though they were the natural product of personalities and circumstances. The shift from "something feels wrong" to "this is the Emotional Extortionist script and here is its countermeasure" is the operational value of classification. It converts the group from a reactive system responding to symptoms to a diagnostic system that can identify the condition and apply the appropriate response.

The synthesis countermeasure is not more sophisticated vetting, tighter surveillance, or heightened suspicion. It is the construction of an environment in which every archetype hits the same wall: a culture organized around orientation to reality rather than charisma, urgency, or personal need; a structure that limits access and damage regardless of trust level; and a detection layer that surfaces orientation drift before it accumulates leverage. That environment is what the Grey Cell Protocols are designed to produce.

What You Can Now Diagnose

- Whether you can identify, for each member who has produced friction, confusion, or operational damage in your system, which archetype's behavioral pattern most closely matches their operating mode, and whether the countermeasure for that archetype has been applied or merely discussed.
- Whether your group's current operating environment provides the Pathological Ally with a crowd that will trade discipline for urgency, the False Prophet with an undocumented narrative that can be gradually reframed, the Emotional Extortionist with a pity leverage point, the Controlled Opposition Plant with an open deliberation space, the Drama Conduit with informal communication channels that operate without discipline, or the Professional Plant with access that is not governed by compartmentalization protocols.
- Whether the archetypes your group is most vulnerable to are the ones your current systems are designed to address, and whether those systems are structural or merely cultural in the sense of being dependent on individual members' vigilance rather than embedded in the group's architecture.

Chapter Notes

1. The six archetypes and their ten-attribute diagnostic framework are drawn from Kit Perez, "Grey Cell Protocols: The Archetypes of Infiltration," The Shepard Scale (Substack), October 5, 2025, https://shepardscale.substack.com/p/gray-cell-protocols-the-archetypes.

2. The property rights organization composite is drawn from observed patterns across multiple civic and advocacy organizational environments. Identifying details have been altered or omitted throughout.

3. The Malheur National Wildlife Refuge standoff, January 2016. For contemporaneous reporting, see Kirk Johnson and Julie

Turkewitz, "Militia Seizes Federal Wildlife Refuge in Oregon," *The New York Times*, January 3, 2016.

4. The legal defense network composite is drawn from observed patterns across multiple volunteer organizational environments. Identifying details have been altered or omitted throughout.

5. The environmental advocacy coalition composite is drawn from observed patterns across multiple advocacy organizational environments. Identifying details have been altered or omitted throughout.

6. The community organizing composite is drawn from observed patterns across multiple organizational environments. Identifying details have been altered or omitted throughout.

7. The animal rights movement infiltration case draws on documented reporting on corporate and law enforcement infiltration of activist organizations. For background, see Will Potter, *Green Is the New Red: An Insider's Account of a Social Movement Under Siege* (San Francisco: City Lights Books, 2011).

CHAPTER 11: COUNTER-NARRATIVE CONTROL

If orientation is the system, narrative is the attack surface. Counter-narrative control keeps the story aligned to reality so the rest of GCP does not ingest corrupted inputs.

Between 1956 and 1971, the FBI's COINTELPRO program targeted more than a hundred domestic organizations: antiwar coalitions, civil rights groups, Black liberation movements, women's collectives, and others.[1] The Bureau did not destroy these groups primarily through surveillance, although there was plenty of that as well. But the real strength of COINTELPRO was in the rewriting of group narratives from within.

Operatives planted stories in newspapers that characterized leaders as informants. Forged letters made allied organizations distrust each other. Inserted agents stoked internal conflict, whispered about betrayals, and distributed rumors calibrated to the specific fracture lines each group already carried. Within a few years, organizations that had been built around shared purpose were consuming themselves. The external threat was real, but the mechanism of destruction was internal.

COINTELPRO's most durable operational insight was that you do not need to defeat a group from outside if you can make it defeat itself from within, and the most reliable way to accomplish that is to corrupt its relationship to its own story.

This is what narrative warfare looks like at scale. It is an orientation operation: a deliberate corruption of the group's relationship to reality, accomplished by corrupting the language and meaning structures through which the group processes reality. Change the story, and you change what members believe is true. Change what members believe is true, and you change what they will do. Force does not become necessary when you have already achieved consent, and consent is manufactured through narrative control.

GCP V is the meaning-integrity layer of the full protocol. GCP I filters through self-disqualification. GCP II replaces emotional trust with structural trust. GCP III reads language and trauma for orientation drift. GCP IV maps the archetypes through which destructive actors operate. GCP V sits above all of them as the system that keeps the story aligned to reality, because if the narrative that feeds the other four facets is corrupted, the inputs to the entire orientation system are corrupted. A group that is running excellent structural vetting on a false story is more dangerous to itself than a group with no vetting at all, because the false story will be defended with the rigor the vetting apparatus provides.

The Four Mechanics of Narrative Hijacking

Narrative hijacking arrives as a series of small adjustments, each of which appears to be a reasonable interpretation of the group's existing commitments. The four mechanics are the standard instruments of that adjustment process.

Pretext manipulation targets the "why" behind an action. The group's mission is rooted in disciplined strategic reasoning. The infiltrator replaces that reasoning with urgency or moral panic. "Real commitment means taking risks." "Do you want to be remembered as cowards or as people who acted when it mattered?" Each statement is framed as an expression of the group's values while simultaneously replacing the group's decision logic with emotional activation.

The mechanism is specific. The infiltrator speaks for the group while speaking to the group, which means the reframing appears to come from inside the group's own value structure rather than from an external source. The action being proposed still looks aligned with the mission's surface features. What has shifted is the underlying motivation: from strategic discipline to emotional self-validation.

Once a group accepts false urgency as a moral good, it stops operating in discipline and begins operating in reaction. The orientation has been moved and the group has not noticed because the language of the mission remained intact while its logic was replaced.

Moral reframing operates at the level of the group's ethical compass rather than its operational logic. The infiltrator does not challenge doctrine directly because direct challenge would expose them. Instead, they shift what the group's values are understood to require. Unity becomes a reason to suppress productive conflict. Operational security becomes reframed as fear: "We aren't going to hide because we aren't afraid," which creates a binary in which anyone who maintains security discipline is implicitly operating from cowardice rather than strategic thinking. Escalation is reframed as inevitability: "We all knew this time would come," which removes the decision from deliberation and positions it as the natural expression of the group's deepest commitments.

These phrases work because they sound noble. They are calibrated to resonate with the group's existing emotional register. The infiltrator needs only to redefine what the group's existing values require. Once that redefinition takes hold, truth stops being the center of the group's orientation. The center becomes emotional resonance: what feels right rather than what is accurate. At that point the group is operating on a corrupted compass, and every decision it makes will compound the initial drift.

Semantic drift is the slowest and most durable of the four mechanics. Where pretext manipulation operates on a single decision and moral reframing operates on the ethical framework, semantic drift operates on language itself: the definitions of the words the group uses to understand its situation and communicate with its members.

The mechanism is cumulative and operates below the threshold of conscious attention. No single shift is large enough to trigger recognition. "Discipline" acquires a connotation of compliance rather than calibrated action. "Brave" begins to mean going along with whatever is proposed regardless of its operational soundness. "Loyal" begins to mean protecting the group's self-description from examination rather than serving its actual mission. Each individual redefinition is small. Their aggregate effect is that the group's language no longer anchors to the same reality it did when the group was formed.

Words are orientation anchors. When they drift, everything attached to them drifts with them. The danger of semantic drift for any doctrine that relies on observation, communication, and reporting is that once the language is corrupted, the data the doctrine generates is corrupted. The same word no longer means the same thing to everyone using it, which means the reports, assessments, and decisions built on that word are built on divergent foundations. Whoever controls the definition controls what the data means, which means they control the narrative without needing to control the facts.

Narrative saturation is the final mechanic and the one that closes the feedback loop permanently if it is not caught. The infiltrator floods the group's communication environment with emotional noise: memes, slogans, inside jokes, half-truths, and emotional appeals calibrated to the group's existing anxieties and commitments. The volume of emotionally activating content exceeds the group's capacity to process contradictory data. When every member is reacting to the same emotional stimuli, the group's feedback system stops functioning as an orientation mechanism and starts functioning as an amplifier for whatever narrative has been saturated into it.

Narrative saturation does not always require a deliberate external actor. Well-intentioned members whose goals are primarily emotional rather than operational spread the same distortion faster than any professional infiltrator because they are genuine, numerous, and have established trust within the group's social architecture. Emotional contagion does not require a source. Once it begins, it self-replicates. By the time saturation is visible as a pattern, the feedback loop has already been compromised, and orientation drift is accelerating without any external intervention required to sustain it.

Why Narrative Sabotage Works: The Four Exploitation Vectors

The mechanics operate through four psychological vulnerabilities that are present in every group regardless of ideological orientation, organizational sophistication, or security culture. Understanding them is mapping the attack surface so that the counter-narrative systems can be designed to address it.

Identity hunger is the human need to belong to a story larger than oneself. For members whose orientation is stable and whose mission commitment is genuine, that need is met through the mission itself, or even through some other internal means such as faith, family, or ethos. For members whose orientation is driven by need rather than reality, the identity hunger is a vacuum.

Infiltrators identify it immediately. They learn who needs to be seen as brave, who needs to feel indispensable, who needs to be recognized as the person who understood what others missed. They offer language that fills those specific needs: "You're the only one who really understands what we're trying to do here." "I can tell you're not the kind of person who backs down." "You were right about this from the beginning."

These statements are pretext. Once a member believes that their emotional value validates the mission, the mission stops being collective and becomes personal theater. Their continued participation is now oriented to the maintenance of the identity the infiltrator has constructed for them. At that point the infiltrator is inside their orientation and can steer it.[2]

Trauma resonance operates through the unresolved wounds that members carry into the group. People who have been controlled crave autonomy. Those who have been unseen crave recognition. Those who have been betrayed crave loyalty above all other values, sometimes above operational accuracy. Infiltrators mirror those wounds. They offer what the trauma seeks: resolution, belonging, recognition, loyalty. The result is dependency that presents as trust. The member does not know they have been captured because the capture feels like finally being understood.

This is why GCP III integrates trauma profiling as a detection tool. The trauma-informed profiles described in Chapter 9 identify exactly the leverage points that narrative hijacking exploits. The shame-hider's secrecy need, the validation addict's recognition need, the approval-seeker's inclusion need: each is a specific vulnerability that the four mechanics are designed to activate. The group that has mapped those vulnerabilities before an infiltrator arrives is significantly harder to compromise than one that discovers them after the damage has been done.

The validation economy is the informal exchange system through which most groups operate without recognizing it. In healthy groups, validation is earned through performance: discipline, reliability, and accurate orientation. In groups whose culture has begun to drift, validation is distributed through emotional alignment: flattery, proximity to the dominant voice, and willingness to echo the prevailing narrative regardless of its accuracy. Infiltrators read that economy immediately and learn who is running a deficit. They print counterfeit currency: emotional affirmation distributed to whoever complies and withheld from whoever challenges. Members begin to self-censor because the social cost of accuracy has become higher than the social cost of agreeing with something they know is wrong.

The institutional version of this dynamic produces the same result through a longer arc. In organizations where emotional compliance has displaced performance as the currency of validation, selection pressure over time concentrates sycophants in influential positions while truth-tellers self-select out, are sidelined, or are formally removed. Remaining members internalize the new rules: sanitize upward information, delay bad news, protect the preferred narrative. The feedback loop collapses through accumulation. Information now confirms whatever the dominant voice requires it to confirm.[3]

Fear of exile is the mechanism that closes the loop. Once dependency has been established and the validation economy has been corrupted, the infiltrator introduces scarcity: the implied or explicit threat that continued belonging requires continued compliance. Members stop challenging dysfunction because they cannot afford to lose their standing in the group.

The group that was formed to resist an external control structure has become the most controllable entity in the room, because its members are now managing their relationship to the group's dominant narrative rather than tracking external reality.

The final stage of narrative sabotage is complete when members silence themselves in defense of a story that no longer exists.

Five Counter-Narrative Systems

The counter-narrative systems do not eliminate the possibility of narrative attack. They make attacks visible before they consolidate by building the group's capacity to test its own story continuously rather than defending it from examination.

Baseline narrative declaration is the first and foundational system. A baseline narrative defines how the group interprets reality: not what it does, but how it decides what to do. It answers three orientation questions explicitly and in documented form. What truth is the group aligned to? What distortions does the group refuse to accept? How does the group measure whether it is still aligned?

The answers to these questions become the orientation anchor against which all subsequent narrative developments are tested. Any reframing of the group's story, however incremental, that cannot be reconciled with the documented baseline is a drift signal. The baseline does not need to be elaborate, just specific enough that departure from it is recognizable.

Narrative audits operationalize the baseline by making its application routine. A narrative audit examines the group's language at regular intervals and asks whether that language still maps to the baseline declaration. Has our language changed? Are new terms or phrases appearing that were not present in the baseline? Are key words being redefined for convenience or pressure? If the way the group talks about its mission is changing faster than the way it is executing that mission, something is steering perception rather than reporting it. The audit frequency should match the group's operational tempo: a group under active pressure may require weekly audits; a group in a stable phase may manage with monthly or quarterly reviews. The mechanism matters more than the frequency.

Challenge culture is the operating environment that makes the first two systems functional. A group that has not institutionalized productive friction cannot run genuine narrative audits because the social cost of surfacing drift is too high. Challenge culture reduces that cost by building dissent into the group's normal operating rhythm: designated devil's advocates in key discussions, anonymous dissent channels where needed, and explicit recognition for members who identify contradictions rather than social consequences for raising them. The crucial distinction is that challenge culture redirects tension toward analysis rather than personal conflict. A group that treats disagreement as orientation work rather than as disloyalty is significantly more resistant to narrative hijacking because the hijacker's tools require an environment in which challenge is costly.

Language discipline treats word choice as observable behavior rather than stylistic preference. What the group permits linguistically, it endorses operationally. The specific fingerprints of narrative infiltration are recognizable in language before they are visible in decisions or actions.

- Urgency without data: "We have to act now."

- Binary framing: "You're either with us or against us."

- Flattery as leverage: "You're the only one who really gets it."

- Redefinition of courage: "We can't hide anymore."

Each of these constructions is a semantic drift event. Training members to recognize them converts language monitoring from a leadership function into a distributed cultural practice. If anyone in the group can recognize and name the fingerprints, then those factors cannot accumulate.

Doctrine reinforcement closes the system by ensuring that the baseline narrative does not become static. Doctrine must be actively revisited and pressure-tested against current operational reality on a regular schedule. Return to foundational statements and mission documents as calibration checkpoints, not as objects of veneration. A story that is continuously updated against verified reality cannot be hijacked by a competing narrative because it has no static version for the competing narrative to displace. Repetition of verified doctrine builds the kind of discipline that semantic drift cannot penetrate because the words are being regularly returned to their anchored meanings rather than allowed to drift between reviews.

Orientation Metrics: Measuring Drift Before Collapse

Groups that rely on intuition to detect narrative drift will consistently miss the inflection point. Drift begins as micro-distortions: a word softened, a motive excused, a challenge deflected without examination. By the time it feels wrong, the feedback loop has already been compromised. The six orientation metrics convert what gut instinct cannot reliably detect into measurable signals that can be tracked over time and compared against baseline.

Alignment consistency measures whether the group's members share a common understanding of what the group is trying to achieve and how it will know when it has succeeded. Administered privately to each member, two questions generate the data: what are we trying to achieve, and how do we know when we have succeeded? If the answers share tone and framework, orientation is stable. If they diverge on motive or meaning rather than on operational detail, fragmentation has begun. Disagreement on implementation is healthy and expected. Disagreement on what the mission means is drift. This metric should be tracked at regular intervals with results compared against prior periods to identify directional movement rather than point-in-time snapshots.

Language drift index monitors the group's communication channels for the substitution of emotional language for procedural language. "We feel called to act" replacing "we verified the conditions." "We're united" replacing "we're ready." "It's time to make a statement" replacing "it's time to execute." Each substitution is a drift event. The metric tracks frequency of substitution and identifies which members are generating it, which reveals whether the drift is distributed across the group or concentrated in specific actors. Concentration in a specific actor is itself a diagnostic signal that warrants archetype examination per Chapter 10.

Validation versus verification ratio measures how the group seeks confirmation. Validation seeks comfort: does the group agree with the proposed course of action? Verification seeks accuracy: is the proposed course of action grounded in reality? A group in which validation requests are outnumbering verification checks has tilted toward emotional orientation. This ratio is the fastest early indicator of manipulation because it detects the shift in the group's epistemology before that shift has produced any specific operational consequence. Healthy groups challenge facts. Drifting groups protect feelings.

Decision latency measures how long the group takes to decide and act on verified information. The metric is diagnostic in both directions. Decisions that arrive too fast indicate urgency pressure, which maps to pretext manipulation from Chapter 10's Pathological Ally or to the pretext manipulation mechanic described earlier in this chapter. Decisions that stall indefinitely indicate ambiguity creation, which maps to the Controlled Opposition Plant. Stable orientation produces predictable decision tempo. Drift produces tempo whiplash: compressed urgency followed by paralysis, followed by compressed urgency again. The rhythm of the group's decision cycle is an orientation indicator that is visible before the content of any specific decision reveals the underlying problem.

Challenge response time tracks how leadership and dominant members respond to legitimate questions and dissent. Fast, fact-based clarification indicates that calibration is intact and the challenge culture is functional. Defensiveness, sarcasm, redirection, or social consequences for the questioner indicate that orientation is being guarded rather than tested. Leadership behavior toward dissent is a more reliable indicator of infiltration risk than any specific content the leadership is protecting, because the pattern of response reveals whether the group's narrative is being treated as an orientation tool or as an identity to be defended.

Drift recovery rate acknowledges that every group drifts and measures how quickly it returns to baseline when drift is identified. This tests how fast it is caught and how efficiently the correction mechanism operates. A group that can identify drift within days and return to baseline within a week is significantly more resilient than one that takes months to recognize and correct the same displacement. Tracking recovery rate over time also reveals whether the group's correction capacity is improving with practice, degrading under pressure, or holding stable, each of which carries different implications for the group's overall orientation health.

Operational Closure

GCP V completes the protocol architecture. The four earlier facets address who is in the group, how much damage any individual actor can do, how to detect orientation drift through language and need, and how to classify the patterns through which destructive actors operate. GCP V addresses the medium through which all those threats are delivered and through which all the system's corrective mechanisms must operate: narrative.

A group that runs GCP I through IV without GCP V is running orientation checks on a story that may already have been corrupted. The data those checks produce will reflect the corrupted story rather than reality, which means the corrections the system generates will compound the drift rather than address it. GCP V ensures that the inputs the system ingests are accurate by making the narrative itself subject to continuous orientation discipline.

Every group that fails, starts that failure the same way: it stops telling the truth to itself. The moment a group protects its story rather than testing it, the battle for orientation is already lost. The counter-narrative systems described in this chapter are offensive discipline: the continuous, institutionalized pressure-testing of the group's story against reality that makes corruption detectable before it consolidates, and correctable before it produces irreversible damage.

Chapter 12 addresses what follows when the full GCP protocol has stabilized orientation: the conditions under which Mission Command can be employed and the relationship between the calibration work GCP represents and the decentralized execution Mission Command requires.

What You Can Now Diagnose

- Whether your group has a documented baseline narrative that new ideas, new members, and new interpretations are tested against, or whether the group's story is held informally in the collective understanding of its founding members and therefore vulnerable to gradual displacement.
- Whether your group runs narrative audits at any regular interval, and whether the people conducting those audits understand what semantic drift looks like in practice before it has accumulated to the point of visibility.
- Whether your group's culture treats challenge as orientation work or as disloyalty, and whether that cultural norm is embedded in the group's operating structure or dependent on the current leadership's tolerance for dissent.
- Whether you can identify, for the last significant conflict or period of dysfunction in your group, which of the four mechanics was operating and which of the four exploitation vectors it was working through, and whether those patterns were visible before they produced their effects.
- Whether your group is currently measuring any of the six orientation metrics, and if not, which one would reveal the most about your group's current orientation state if it were measured today.

Chapter Notes

1. COINTELPRO program documentation and historical record: U.S. Senate, Select Committee to Study Governmental Operations with Respect to Intelligence Activities (Church Committee), Final Report, Book III: "Supplementary Detailed Staff Reports on Intelligence Activities and the Rights of Americans" (Washington, DC: U.S. Government Printing Office, 1976). For secondary analysis of COINTELPRO's narrative and organizational disruption tactics, see Ward Churchill and Jim Vander Wall, *The*

COINTELPRO Papers: Documents from the FBI's Secret Wars Against Dissent in the United States (Boston: South End Press, 1990).

2. The counter-narrative framework, including the four mechanics of narrative hijacking, the four exploitation vectors, and the five counter-narrative systems, is drawn from Kit Perez, "Grey Cell Protocols: Counter-Narrative Control," The Shepard Scale (Substack), October 12, 2025, https://shepardscale.substack.com/p/grey-cell-protocols-counter-narrative.

3. For the institutional expression of this dynamic at organizational scale, including the selection pressure that concentrates sycophants in positions of influence while forcing truth-tellers to self-select out, see Chapter 7 of this volume. See also Donald E. Vandergriff, *Adopting Mission Command: Developing Leaders for a Superior Command Culture* (Annapolis, MD: Naval Institute Press, 2019).

CHAPTER 12: CALIBRATION BEFORE DECENTRALIZATION

Anytime orientation has been stabilized, GCP must allow Mission Command to emerge.

The argument of this book has been structural: Mission Command fails when the conditions those principles require are absent but commanders attempt to employ them. Shared orientation, structural trust, psycholinguistic awareness, archetype recognition, narrative integrity: these are Mission Command's prerequisites.

A decentralized execution system deployed into a group that lacks stable orientation, structural trust, and narrative integrity produces accelerated Mission Command failure, because decentralization amplifies whatever the system already contains, and a system that contains misalignment, narrative corruption, or infiltration will express those conditions faster and at greater scale when the controls that were limiting them are removed.

The Grey Cell Protocols address this problem sequentially. GCP I through V establish the conditions that Mission Command requires before decentralized execution is attempted. That sequencing is the restoration of what Mission Command doctrine has always assumed: that the trust, orientation, and cultural foundations it depends on exist before the authority to act on them is distributed downward.

Don Vandergriff's five-action restoration sequence makes this sequencing explicit in operational terms.[1] The sequence is:

- rebuild trust through personal vulnerability

- protect truth-tellers

- restore healthy embarrassment through calibration ritual

- decentralize execution by issuing intent rather than orders

- institutionalize learning through facilitated after-action reviews.

Action 4, decentralized execution, is fourth in a sequence of five because decentralization without the first three actions produces exactly the conditions Part II of this book described. Trust that has not been rebuilt through demonstrated vulnerability will not sustain the exposure that decentralized execution requires. Truth-tellers who have not been structurally protected will not surface the orientation data that decentralized execution depends on. Calibration rituals that have not been institutionalized will not catch the orientation drift that decentralized execution accelerates when it occurs.

The Grey Cell Protocols map onto this sequence as the calibration infrastructure that Actions 1 through 3 require and that Action 4 presupposes.

- GCP I builds the self-disqualifying environment that makes trust structurally grounded rather than personally dependent.
- GCP II establishes the damage containment architecture that makes trust survivable when it is tested.
- GCP III provides detection tools that make calibration operational rather than aspirational.
- GCP IV names the patterns that calibration must be protected against.
- GCP V maintains the narrative integrity that calibration requires to produce accurate outputs.

Together they constitute the calibration phase. Action 4 is the exploitation phase. Action 5 is the maintenance mechanism that keeps the exploitation phase from degrading back into conditions that require recalibration.

GCP as Calibration Phase

The calibration phase has a specific operational objective: stable orientation. Not perfect orientation, or ironclad immunity to drift, infiltration, or narrative attack (although these can be nearly reached in practice with intentionality and discipline). Stable orientation means that the group's decision inputs are tracking reality with sufficient accuracy that decentralized execution will produce results that serve the mission rather than compound existing misalignment.

Stable orientation is measurable. The seven metrics Vandergriff identifies for restoration progress apply directly as calibration readiness indicators: [2]

- candor index
- initiative events
- truth-teller protection rate
- tempo delta
- trust score
- misalignment admission rate
- follow-through rate.

A group that scores adequately across these metrics has demonstrated that its orientation is stable enough to sustain decentralized execution. A group that cannot sustain a candor index above threshold, that has not protected truth-tellers consistently, or whose trust scores remain below the confidence threshold has not completed the calibration phase regardless of how long GCP has been running.

The calibration phase ends when the conditions it was designed to establish are demonstrably present. It is a system state: the group is oriented to reality, its trust architecture is structural rather than personal, its detection tools are active and producing accurate data, its members understand and can apply the archetype framework, and its narrative is anchored to a documented baseline that is being regularly audited. When that state is demonstrably present, the calibration phase is complete.

This means GCP has a natural termination point within any given operational cycle, which is a feature rather than a limitation.

GCP is designed to make itself unnecessary as a primary operating mode by establishing the conditions under which Mission Command can be employed without the failures Part I of this book documented.

Mission Command as Exploitation Phase

When orientation is stable, Mission Command is not optional. It is the operationally correct response to the conditions that GCP has established. A commander who has run GCP to stable orientation and then continues to operate through GCP's vetting and verification mechanisms rather than transitioning to decentralized execution is squandering the conditions they have built. They are also signaling to their subordinates that demonstrated trustworthiness is not sufficient to earn operational authority, which is precisely the message that erodes the trust GCP spent its calibration phase constructing.

The transition to Mission Command requires commander's intent issued at the appropriate level of specificity: purpose, key tasks, end state, and acceptable risk parameters. It requires subordinates who have been briefed on intent and whose understanding of it has been verified before execution. It requires the commander's active restraint from method-level intervention: adjusting resources and end state when necessary, not directing how subordinates achieve what they have been tasked to achieve. And it requires the after-action review culture that Action 5 establishes: the institutionalized learning mechanism that catches drift in the exploitation phase before it requires a full return to calibration.

An Army company-level element demonstrates a successful transition from GCP to Mission Command. They had spent approximately eight months in what its commander recognized as a calibration phase: rebuilding candor through structured AARs, replacing two members whose orientation drift had been identified through psycholinguistic monitoring, addressing a narrative saturation event that had temporarily disrupted the unit's operational focus, and establishing compartmentalized access protocols that removed the single points of failure that had existed in its previous personnel structure. By the end of that period, the unit's candor index was consistently above threshold, truth-tellers had been protected through two significant incidents, and the commander's trust score reflected genuine confidence rather than social compliance.

The transition to Mission Command was not announced as a policy change. The commander issued intent for the next operational cycle at a level of specificity that assumed subordinate competence rather than requiring subordinate compliance. He intervened in execution only twice over the following few months, both times to adjust resources rather than method.

The unit's tempo accelerated. The orientation accuracy that the calibration phase had established was now producing faster, more adaptive decision cycles because the actors making decisions at the point of execution had the orientation stability to make them well. The calibration phase had done its work. The exploitation phase was expressing that work in operational outputs.[3]

Recalibration Without Ego: Returning to GCP Is Routine Maintenance

Just like GCP is never meant to be a permanent state, the transition to Mission Command is not permanent and is not intended to be. Orientation still degrades under pressure and needs to be recalibrated consistently. New members enter the system with unknown orientation states. External actors attempt narrative hijacking regardless of the group's internal health. Events generate stressors that activate the shame, validation, and approval-seeking dynamics that GCP III identified as orientation vulnerabilities.

A system that transitions to Mission Command and then treats the calibration tools as retired is a system that will eventually require emergency recalibration rather than routine maintenance.

The correct model is cyclical rather than linear. Calibration phase establishes stable orientation. Exploitation phase decentralizes execution against that stable foundation. Maintenance phase monitors orientation metrics, runs narrative audits, applies archetype detection tools when new actors enter the system, and catches drift before it reaches the threshold requiring a full return to calibration. When maintenance catches drift that exceeds the maintenance phase's corrective capacity, the system returns to calibration mode for the affected subsystem or the full group depending on the scope of the drift. That return is the system functioning as designed.

The ego risk at this juncture is significant and worth naming directly. A commander or leader who has successfully transitioned to Mission Command may resist returning to calibration mode because the return feels like an admission that the transition was premature or that the exploitation phase failed. Neither interpretation is accurate.

Orientation degrades because systems are prone to entropy. The conditions that were stable when the transition was made are not guaranteed to remain stable indefinitely. A leader who recognizes orientation drift and initiates recalibration is exercising exactly the kind of accurate self-assessment that the calibration phase was designed to build. A commander who resists recalibration because it feels like regression is protecting a narrative rather than tracking reality, which is the mechanism Chapter 11 identified as the beginning of the end for any group that pursues it.

An organization that had built strong orientation discipline over two years and was operating effectively under distributed authority experienced a significant personnel transition: three founding members departed within a four-month period, each of whom had been carrying substantial informal orientation infrastructure that had never been formally documented or distributed. The incoming members were capable and committed, but their orientation to the group's mission, culture, and operational norms had not been established through a calibration phase equivalent to what the founding members had experienced.[4]

The organization's leadership recognized the departure pattern but did not return to calibration mode. The exploitation phase continued. Within six months, two of the orientation metrics that the organization had previously used to monitor system health had degraded significantly: the candor index had dropped below threshold and the validation versus verification ratio had tilted toward validation. The narrative drift that followed was not dramatic. It was gradual and cumulative, exactly the pattern Chapter 11's semantic drift mechanic described. By the time the degradation was recognized as requiring intervention, the organization had lost eight months of operational capacity to a problem that a return to calibration mode at the point of the personnel transition would have addressed in four to six weeks.

The lesson is operational rather than cautionary. Recalibration triggered by a significant system change, whether personnel transition, external pressure, or detected orientation drift, is cheaper and faster than emergency recalibration triggered by system failure. A leader who builds the habit of monitoring orientation metrics continuously and initiates calibration responses at the maintenance threshold rather than the crisis threshold is displaying exactly the operational discipline that Mission Command requires of the leaders it trusts with decentralized authority.

Operational Closure

The Grey Cell Protocols and Mission Command are sequential phases of a single operational approach to command under uncertainty. GCP establishes the conditions. Mission Command exploits them. The sequencing is not optional. The exploitation phase deployed without the calibration phase produces the failure modes documented in Parts I and II of this book. The calibration phase that never transitions to the exploitation phase produces control without capacity, which is Mission Command's opposite and its enemy.

Chapter 13 addresses the diagnostic questions that bridge the two phases:

- How can you correctly assess which phase your system currently requires?
- What are the indicators of calibration readiness?
- What are the consequences of assessing incorrectly?

What You Can Now Diagnose

- Whether the transition from GCP to Mission Command in your system, if it has occurred, was triggered by demonstrated orientation stability or by timeline pressure, leadership preference, or the assumption that sufficient time had passed.
- Whether your system has a recalibration trigger: a defined threshold at which orientation metric degradation initiates a return to calibration mode.
- Whether your commanders understand that returning to calibration mode is a function of accurate orientation rather than an admission of failure, and whether that understanding is embedded in the organizational culture or dependent on the current leadership's willingness to model it.

Chapter Notes

1. The five-action restoration sequence is drawn from Donald E. Vandergriff, *Adopting Mission Command: Developing Leaders for a Superior Command Culture* (Annapolis, MD: Naval Institute Press, 2019). The sequence as described: rebuild trust through personal vulnerability, protect truth-tellers, restore healthy embarrassment through calibration ritual, decentralize execution by issuing intent rather than orders, and institutionalize learning through facilitated after-action reviews.

2.   The seven orientation metrics are drawn from Vandergriff, *Adopting Mission Command.* The candor index, initiative events, truth-teller protection rate, tempo delta, trust score, misalignment admission rate, and follow-through rate are described as observable, non-bureaucratic behavioral indicators tied directly to orientation health.

3.   The company-level transition composite is drawn from observed patterns in military leader development contexts documented in Vandergriff, Adopting Mission Command, and the authors' combined observations across military and civilian leadership development environments. Identifying details have been omitted throughout.

4.   The civic organization recalibration composite is drawn from observed patterns across multiple organizational environments. Identifying details have been altered or omitted throughout.

CHAPTER 13: DIAGNOSING THE SYSTEM STATE

Effective command depends on correctly assessing system orientation and applying the appropriate framework. Misdiagnosis in either direction guarantees collapse.

Every framework in this book presupposes a prior act of diagnosis. Mission Command presupposes that orientation is stable, trust is structural, and subordinates are calibrated to the commander's intent. The Grey Cell Protocols presuppose that those conditions are absent or degraded and require active calibration. Neither framework is universally applicable or a default. The question that precedes every command decision is not which framework to prefer but *which framework the system's current state requires.*

That question is harder than it sounds. Leaders who have read this far understand the conditions Mission Command requires and the tools GCP provides to establish them. What is less obvious is that both frameworks can be misapplied with equal confidence and equal damage.

A commander who deploys Mission Command into a system that lacks stable orientation does not produce decentralized excellence. A commander who runs GCP indefinitely in a system that has achieved stable orientation does not produce security. Both misdiagnoses produce collapse, and both of these mistakes are common precisely because the pressure to apply one framework or the other rarely comes with accurate data about the system's actual state.

This chapter provides the diagnostic framework to know the difference. The indicators are organized into two sets: conditions that require GCP and conditions that enable Mission Command. Misdiagnosis in either direction is then examined as a specific failure mode with predictable consequences.

Indicators Requiring GCP

The indicators requiring GCP are system-level signals that the orientation infrastructure Mission Command depends on is absent, degraded, or actively under attack. A single indicator in isolation warrants attention. Multiple indicators presenting simultaneously warrant immediate diagnosis and GCP activation regardless of how long the system has been operating, or how stable it appeared previously.

The first and most fundamental indicator is **candor failure**. When subordinates consistently sanitize reports, when bad news arrives late or not at all, when after-action reviews produce consensus rather than contradiction, the system's feedback loop has been compromised.

John Boyd's insight that orientation is where accurate maps of reality are built or distorted applies directly here.[1] A system that cannot surface accurate information cannot orient accurately, and a system that cannot orient accurately cannot decentralize execution without amplifying whatever distortions it is already carrying. Candor failure is an orientation system failure that makes Mission Command operationally dangerous.

The second indicator is **trust that is personal rather than structural**. When the system's cohesion is organized around specific individuals rather than around documented architecture, the conditions of Chapter 8 apply: the system is one departure or one betrayal away from losing the informal infrastructure it has been depending on.

Personal trust is not inherently problematic. It becomes a GCP indicator when it is the primary or exclusive trust mechanism, when access and authority are distributed based on relationship rather than protocol, and when the departure or compromise of a key individual would produce damage disproportionate to what the system's structure could contain. A system organized around personal trust has not completed GCP II, regardless of how stable it feels.

The third indicator is orientation drift that is visible in language before behavior. Semantic drift in the group's communication channels, urgency framing appearing without verified data to support it, the validation versus verification ratio tilting toward validation: these are the early signals Chapter 11 identified as detectable before they produce operational consequences.

When these signals are present, the narrative integrity layer has been breached or is under active pressure. Decentralizing execution into a system whose narrative is drifting distributes the drift rather than containing it.

The fourth indicator is the presence of undiagnosed archetypes. When friction, stagnation, or accelerating dysfunction cannot be attributed to operational factors and the system has not applied Chapter 10's diagnostic framework to its current membership, the possibility that one or more archetypes are operating inside the system must be treated as an open diagnostic question rather than a resolved one. A system that has not run archetype diagnosis on its current personnel does not know what it is decentralizing into.

The fifth indicator is new personnel whose orientation has not been established through a calibration process equivalent to what the existing members experienced. Personnel transitions are the most common trigger for orientation regression in systems that have previously achieved stability, as the composite case in Chapter 12 demonstrated. When new members enter a system operating in exploitation phase, they bring unknown orientation states into an environment whose architecture was calibrated around the people who built it.

The appropriate response is to run GCP on the incoming members as a discrete calibration process before integrating them into decentralized execution roles. Imagine the Navy integrating a few new members onto the SEAL teams without them going through BUD/S, and you will understand this indicator.

The sixth indicator is recovery failure: the system has identified orientation drift but its correction mechanisms have not returned it to baseline within a timeframe proportionate to the drift's severity. A system that can identify drift but cannot correct it has a structural problem in its calibration architecture that GCP must address before Mission Command can be safely maintained.

Indicators Enabling Mission Command

The indicators enabling Mission Command are the measurable expressions of stable orientation. These do not need to be perfect; indeed, perfection is something strived for and never attained here. The goal is demonstrated sufficiency: the system is oriented to reality with enough accuracy and structural integrity that decentralized execution will produce mission-aligned outputs rather than compounding existing misalignment.

The candor index is above threshold and has been sustained across multiple cycles. Subordinates are surfacing bad news early, challenging flawed assumptions, and receiving protection rather than consequences for doing so. The feedback loop is producing accurate data. Observation is not being curated to protect preferred narratives.

Structural trust is the operating architecture. Access and authority are distributed by protocol rather than relationship. The damage any single actor can produce is limited by compartmentalization and redundancy rather than by confidence in that actor's continued reliability. The system has been tested against at least one significant personnel change or trust stress and has demonstrated that its architecture contained the damage within designed parameters.

The validation versus verification ratio favors verification. Members are seeking accuracy rather than comfort when they need confirmation. The group's decision culture is organized around what is true rather than around what will be affirmed. Language discipline is active and distributed: members can identify the fingerprints of narrative infiltration and surface them without social cost.

The archetype framework has been applied to current personnel, and no active archetype has been identified, or where one was identified it has been neutralized through the appropriate countermeasure. The system is not operating in a state of undiagnosed threat.

Decision latency is stable and proportionate to operational tempo. The system is not experiencing urgency pressure that is compressing decisions beyond the calibration the orientation infrastructure can support, and it is not experiencing the paralysis that indicates controlled opposition is operating in its deliberation space.

Commander's intent can be issued at the appropriate level of specificity and subordinates can brief back their understanding of it accurately before execution. This is the operational test of shared orientation: not whether the commander believes subordinates are aligned but whether subordinates can demonstrate alignment in their own words against the commander's standard before they are released to act on it.

When these conditions are demonstrably present across multiple measured cycles rather than inferred from general impressions, the system is in a state that Mission Command requires the commander to exploit. The conditions have been established. The infrastructure is in place. Continued GCP operation beyond this point is over-control of a calibrated system, and over-control has its own failure mode.

Consequences of Misdiagnosis

Misdiagnosis in either direction produces a specific failure mode with predictable operational consequences.

Deploying Mission Command into unstable orientation is the more commonly recognized failure mode and the one Parts I and II of this book documented at length. When decentralized execution is attempted in a system whose orientation is degraded, the decentralization distributes the degradation. Subordinates acting on intent they have not accurately received produce actions misaligned with the commander's actual purpose. The False Prophet's narrative drift, operating in a system without the counter-narrative controls of GCP V, propagates faster under decentralized conditions because there is no centralized check on the outputs the drift is producing. The Pathological Ally's urgency pressure finds a wider audience when subordinate commanders have authority to act on it. The Professional Plant's structural compromise extends further when compartmentalization has not been established before authority is distributed.

The operational signature of this misdiagnosis is initial apparent success followed by accelerating incoherence. The system appears to be functioning because the energy of decentralized action is visible and generates the appearance of tempo. The incoherence accumulates beneath that surface because the actions being taken are not converging on the commander's actual intent. By the time the incoherence is visible at the command level, the system has already absorbed costs that a correct diagnosis would have prevented.[2]

Boyd's warning applies with precision here: mismatched orientation leads to incurable rigidity, because the distorted map it is acting on cannot be corrected by the outputs the map is generating.[3] The system that has decentralized into unstable orientation is producing data that confirms its distortion rather than correcting it, because the actors generating that data are operating from the same distorted map the system deployed them with.

Maintaining GCP beyond demonstrated orientation stability is the less commonly recognized failure mode and the more insidious one, because it presents as caution rather than error. A commander who has run GCP to stable orientation and continues operating in calibration mode is signaling to subordinates that demonstrated trustworthiness is insufficient to earn operational authority. That signal erodes precisely the trust the calibration phase built.

The operational signature of this misdiagnosis is progressive disengagement by the system's most capable members. High-agency subordinates who have demonstrated orientation stability and earned the authority to act on their calibrated judgment will not indefinitely sustain engagement in a system that withholds that authority indefinitely. They either disengage internally, reducing their outputs to the minimum the control system requires, or they depart for environments that will use what they have built. Either outcome degrades the system's operational capacity by exactly the margin that its most capable members represent.

The second consequence is that continued GCP operation in a stable system consumes the calibration resources that maintenance-level orientation monitoring requires.A system running full calibration protocols on a stable population is spending calibration capacity on verification of conditions that have already been verified, which leaves less capacity available for the monitoring that will detect the next genuine drift event. The system that cannot distinguish between calibration mode and maintenance mode will eventually fail to detect the drift event that requires intervention because its detection resources are fully committed to verifying stability that has already been established.

The third consequence is doctrinal. GCP is a calibration tool. A system that treats it as a permanent operating mode has misunderstood what it is. The purpose of the Grey Cell Protocols is to establish the conditions Mission Command requires and then yield to Mission Command. A system that runs GCP indefinitely has not completed the doctrine. It has stopped at the prerequisite and mistaken it for the destination.

The Diagnostic Practice

Correct diagnosis requires that the indicators described above be assessed against measured data rather than general impressions. A leader who believes their system is oriented accurately because it feels cohesive, because members appear committed, or because no obvious disruption has occurred is assuming. The orientation metrics of Chapter 11, the calibration readiness indicators of Chapter 12, and the archetype diagnostic framework of Chapter 10 exist precisely because the felt sense of a system's health is an unreliable diagnostic instrument.

Systems feel healthy until they do not, and the transition from feeling healthy to visibly failing is the period during which accurate measurement would have produced actionable data.

The diagnostic practice is therefore a continuous rhythm.

- Alignment consistency measured at regular intervals.
- Language drift index monitored against baseline.
- Validation versus verification ratio tracked across decision cycles.
- Challenge response time observed in the system's actual behavior rather than in its stated norms.

Full definitions, measurement methods, and diagnostic thresholds for all thirteen orientation metrics referenced in this book are provided in Appendix A. These measurements, applied consistently and compared against prior periods, produce the data that correct diagnosis requires.

A leader who has established this rhythm does not need to ask whether the system is ready for Mission Command or whether it requires GCP. The data answers that question. The commander's role is to read the data accurately and act on what it says rather than on what they would prefer it to say. That act of accurate reading, without ego protection and without preference for a particular answer, is the orientation discipline this entire book has been building toward.

It is also the discipline that the Grey Cell Protocols were designed to support: not as a permanent operating mode, not as a substitute for command judgment, but as a structured means of ensuring that the commander's map of their system's state is tracking reality rather than confirming a preferred narrative about it.

That is what orientation-based command looks like in practice. And that is what this book has been arguing, from the first chapter to this one.

What You Can Now Diagnose

- Whether your system's current state matches the framework you are currently applying to it.
- Whether you can identify, for each of the six GCP indicators described in this chapter, the current measured state of your system against that indicator, and whether the aggregate of those measurements points toward calibration phase, exploitation phase, or maintenance phase.
- Whether your system has experienced the over-control failure mode: capable subordinates disengaging or departing from a system that achieved orientation stability but did not transition to decentralized execution in time to retain them.
- Whether your commanders understand the diagnostic practice as a rhythm rather than an event, and whether that rhythm is embedded in the system's operating structure or dependent on individual leaders choosing to apply it.
- Whether you can read your system's current data and state clearly which phase it is in and why, in terms specific enough that another commander reading the same data would reach the same conclusion.

Chapter Notes

1. John R. Boyd, "Destruction and Creation" (unpublished paper, September 3, 1976); John R. Boyd, "The Essence of Winning and Losing" (unpublished briefing, January 1996). Boyd's formulation of orientation as the phase where genetic heritage, cultural traditions, previous experience, and ongoing analysis are

synthesized into an implicit map of reality is the foundational theoretical basis for the orientation-based approach developed in this volume. For extended treatment of Boyd's framework in military leadership context, see Donald E. Vandergriff, *Adopting Mission Command: Developing Leaders for a Superior Command Culture* (Annapolis, MD: Naval Institute Press, 2019).

2. The operational signature of Mission Command deployed into unstable orientation is drawn from composite cases documented in Vandergriff, Adopting Mission Command, and from the authors' combined observations across military and civilian leadership development environments.

3. Boyd, "The Essence of Winning and Losing." The formulation that mismatched orientation produces incurable rigidity rather than flexible response is central to Boyd's argument about why orientation is the decisive phase of the OODA loop.

CONCLUSION: COMMAND WITHOUT DELUSION

Mission Command is a conditional system state that emerges only when prerequisite conditions are structurally established and maintained. The Grey Cell Protocols provide the infrastructure required to do so.

The argument of this book is not complicated: Mission Command works when its prerequisites are in place and fails when they are not.

The failures documented in Part I are failures of the conditions the doctrine assumes. Shared orientation, structural trust, calibrated feedback, and narrative integrity do not emerge from declaring Mission Command as organizational philosophy. They must be built, measured, and maintained as a prior discipline. Without that discipline, decentralized execution distributes whatever the system already contains, and a system that contains misalignment, infiltration, or narrative corruption will express those conditions faster and at greater scale when the controls limiting them are removed.

This is not a new insight in its parts. Boyd established that orientation is the decisive phase of the decision cycle and that a distorted map of reality produces decisions that compound the distortion rather than correct it. Vandergriff documented the institutional mechanisms that have systematically degraded orientation in military leadership culture and developed the restoration methodology required to reverse that degradation. Perez's Grey Cell Protocols established orientation-based vetting as a structured doctrine applicable across organizational contexts, with specific tools for self-disqualification, structural trust, psycholinguistic detection, archetype recognition, and narrative integrity.

What this book has done, for the first time, is formalize the sequential relationship between those two bodies of work. GCP is the calibration phase. Mission Command is the exploitation phase.

That sequencing is the doctrinal contribution this volume makes, and we state it precisely: this book is the first to formally integrate the Grey Cell Protocols as the structured prerequisite infrastructure for Mission Command employment and to specify the diagnostic criteria that determine readiness for that transition. The components of this synthesis exist independently in prior literature. The integrated doctrine does not.

What the Synthesis Produces

The synthesis is not additive; GCP and Mission Command are not two parallel frameworks a leader can choose between based on preference or context. It is a single integrated doctrine with two sequential phases and a diagnostic layer that determines which phase applies at any given moment.

The calibration phase does specific work. GCP I builds the self-disqualifying environment that filters misaligned actors before they consolidate. GCP II establishes the structural trust architecture that limits the damage any actor can produce regardless of how long they have been present or how much trust they have accumulated. GCP III provides the psycholinguistic and trauma-informed tools that surface orientation drift through language before it produces behavioral consequences. GCP IV names the archetypes through which destructive actors operate, converting diffuse suspicion into actionable diagnosis. GCP V maintains the narrative integrity layer that keeps the story the system is telling itself aligned to reality rather than to a preferred version of it.

The exploitation phase uses what the calibration phase built. Commander's intent issued at the appropriate level of specificity. Subordinates who have demonstrated orientation stability exercising disciplined initiative within that intent. A feedback architecture that surfaces bad news early, protects the people who surface it, and closes the loop between identified misalignment and implemented correction. Decision cycles that operate at the tempo the operational situation demands, because the orientation foundation supporting them is stable enough to sustain that tempo without compressing the accuracy the decisions require.

The diagnostic layer reads the system's current state against measured indicators and determines which phase applies. It relies purely on data: the thirteen orientation metrics of Appendix A, assessed at regular intervals and compared against documented baseline, producing an answer that the system's actual state generates rather than an answer the commander's preference selects.

Together these three elements produce what the title of this conclusion names: command without delusion.

- Command that is oriented to the system's actual state rather than to a preferred narrative about it.
- Command that distributes authority when the conditions for distributing it exist and withholds authority when they do not.
- Command that returns to calibration when the exploitation phase has exceeded its sustainable conditions, without ego protection and without the misreading of that return as regression.

The Failure Modes This Doctrine Prevents

A leader who has absorbed this book can now recognize, before they produce irreversible damage, the two failure modes that have ended more organizations than any external threat.

The first is deploying decentralized execution into a system whose orientation has not been established. The energy of distributed action generates the appearance of capability. The incoherence accumulates beneath it. By the time it is visible at the command level, the system has absorbed costs that correct diagnosis would have prevented. The leader who cannot distinguish between orientation stability and social cohesion, between structural trust and personal loyalty, between a calibrated feedback loop and a candid-seeming conversation, will misread the system's state and deploy Mission Command into conditions that will break it. This book provides the diagnostic tools that prevent that misreading.

The second is maintaining calibration controls in a system that has achieved orientation stability. Capable subordinates who have demonstrated accurate orientation and earned the authority to act on it will not sustain indefinite engagement in a system that withholds that authority. They disengage or depart. The system loses exactly the margin that its most capable members represent, which is typically the margin that determines whether it succeeds or fails under operational pressure.

The leader who mistakes continued GCP operation for caution rather than recognizing it as over-control will exhaust the system's best people at the moment they are most needed. This book names that failure mode explicitly so that it can be recognized before it completes.

Both failure modes share a common root: the commander's map of the system's state has detached from the system's actual state. The commander is making decisions about which framework to apply based on a distorted orientation rather than on what the measured indicators say. Boyd's warning applies in full: mismatched orientation produces decisions that compound the distortion. The Grey Cell Protocols and the diagnostic framework of Chapter 13 exist to prevent that compounding by keeping the commander's map of their system calibrated to the terrain the system is navigating.

The Stakes

Superficial reform does not address this problem. Another training module, another climate survey, another leadership development program that does not change the structural incentives producing the failure modes it claims to address: none of these produce the reorientation the problem requires.

Every institution this book has examined had access to the principles it was violating. The problem is the absence of a structured discipline for establishing and maintaining the conditions those principles require.

The stakes in military contexts are existential in the precise sense of the word. Future conflict will be defined by speed, uncertainty, and information saturation. Adversaries will operate with flatter structures, greater tolerance for risk, and the explicit intention of exploiting the decision latency that centralized, ego-protective, narrative-managed institutions produce. A force that has prioritized optics over outcomes, compliance over initiative, and personal loyalty over structural trust will be outcycled by an adversary that has not. Human lives will pay the cost of that misalignment.

The stakes in civilian contexts are proportionate to what the organization is trying to accomplish and who depends on it. A civic organization that collapses under infiltration it could have detected forfeits whatever ground it held against whatever it was opposing. A leadership team that cannot distribute authority because it has not established the conditions for doing so leaves capability on the table at exactly the moment the situation requires it.

Orientation is simply the system's map of reality. A system whose map is accurate will make better decisions than a system whose map is distorted, regardless of how committed, capable, or well-resourced either system is. That is the foundational claim of every chapter in this book, and it is the claim that the synthesis of Mission Command and the Grey Cell Protocols is designed to make actionable.

What You Can Now Do

Before you chose to read this book, you could probably observe that your organization was not functioning as it should. You could sense that trust was fragile, that decisions were slow, that the people most capable of accurate assessment were not the people whose assessments were reaching the decision cycle. You could feel the friction and name its symptoms. What you could not do was convert that observation into a structured diagnosis, apply a specific countermeasure to the specific failure mode the diagnosis identified, and measure whether the countermeasure was producing the conditions it was designed to produce.

You can do that now.

You can assess your system's current state against the indicators of Chapter 13 and determine whether it requires calibration or is ready for exploitation. You can apply the self-disqualifying environment of GCP I before a destructive actor consolidates rather than after they have produced their damage. You can design the structural trust architecture of GCP II so that the cost of any individual's failure or betrayal is contained within parameters the system can absorb. You can apply GCP III's psycholinguistic framework to surface orientation drift through language before it becomes behavioral. You can name the archetype when you see the pattern and apply the specific countermeasure rather than managing a vague sense that something is wrong. You can run the narrative audit that catches semantic drift before it has displaced the baseline story the system was built on.

And when the calibration phase has done its work, when the indicators are above threshold and the system is demonstrably oriented to reality, you can issue intent at the appropriate level of specificity, release your subordinates to act on it, and trust that the architecture you have built will surface the misalignments that require correction before they compound into failures that require reconstruction.

That is command without delusion. It is not a permanent state; it requires continuous diagnosis, periodic recalibration, and the discipline to read the system's actual state rather than the state you would prefer it to be in. But it is achievable. We have seen it. And this book has given you the doctrine to build it.

The drift is deep, but the choice is yours.

APPENDIX A: ORIENTATION METRICS REFERENCE

This appendix provides operational definitions, measurement methods, target thresholds, and diagnostic implications for the thirteen orientation metrics used throughout this book.

The metrics are organized into two sets. The GCP Metrics (1--6, plus Drift Recovery Rate) are drawn from the Counter-Narrative Control framework developed in Chapter 11 and apply to any organizational environment in which orientation health is being monitored.

The Restoration Metrics (7--13) are drawn from Vandergriff's five-action restoration sequence and apply specifically to systems in active calibration or assessing readiness to transition to Mission Command.

Both sets measure the same underlying system state from different angles. Used together, they provide a complete diagnostic picture. No single metric is conclusive in isolation. Patterns across multiple metrics are the reliable signal.

GCP Orientation Metrics

1. Alignment Consistency

What it measures: Whether members share a common understanding of what the group is trying to achieve and how it will know when it has succeeded.

How to measure it: Administer privately to each member, individually and without group discussion, two questions: What are we trying to achieve? How do we know when we have succeeded? Compare answers for convergence of tone, framework, and stated motivation. Divergence on operational detail is normal. Divergence on meaning, purpose, or what success requires is a drift signal.

Target threshold: Answers should share consistent framework and motivation across at least eighty percent of members. Semantic spread between members on core purpose warrants immediate investigation.

Diagnostic implication: Fragmentation in alignment consistency indicates that the group is operating from divergent maps. Decentralized execution under these conditions will produce divergent outputs. Run full GCP III psycholinguistic review on members showing significant divergence before proceeding.

Frequency: Quarterly at minimum. Monthly during high-pressure periods or following significant personnel changes.

2. Language Drift Index

What it measures: The rate at which emotional language is replacing procedural language in the group's communication channels.

How to measure it: Audit meeting notes, written communications, and recorded discussions over a defined period. Count substitution events: instances in which procedural language has been replaced by emotional framing. Examples include "we feel called to act" replacing "we verified the conditions," "we're united" replacing "we're ready," or "it's time to make a statement" replacing "it's time to execute." Track frequency of substitution events per audit period and identify which members are generating them.

Target threshold: Zero substitution events is the ideal. Any substitution warrants notation. A pattern of substitution across multiple audit periods, or substitution events concentrated in a single member, warrants archetype diagnosis per Chapter 10.

Diagnostic implication: Language drift precedes behavioral drift. A rising language drift index is an early warning that the narrative integrity layer is under pressure. Concentration in a specific member may indicate False Prophet or Drama Conduit activity.

Frequency: Monthly audit of communication channels. Real-time monitoring during high-operational-tempo periods.

3. Validation vs. Verification Ratio

What it measures: The balance between comfort-seeking and accuracy-seeking in how members request confirmation.

How to measure it: In meetings, discussions, and written exchanges, classify confirmation requests by type. Validation requests seek agreement or affirmation: "Do you think this is the right call?" "Does everyone feel good about this?" Verification requests seek accuracy: "Is this information confirmed?" "What's the source for that assessment?" Track the ratio of validation to verification requests across a defined observation period.

Target threshold: Verification requests should outnumber or equal validation requests. A ratio tilting significantly toward validation indicates the group's epistemology has shifted toward emotional orientation.

Diagnostic implication: This is the fastest early indicator of manipulation. When the ratio tilts toward validation, the group is protecting feelings rather than tracking facts. The shift will precede visible behavioral consequences by weeks to months. Initiate narrative audit and GCP III review when the ratio moves beyond baseline.

Frequency: Continuous observation. Formalize the count quarterly.

4. Decision Latency

What it measures: The time between verified information becoming available and a decision being made and executed on that information.

How to measure it: Track the interval between the point at which actionable information is available to the decision-maker and the point at which a decision is made and communicated. Compare against the group's established operational tempo baseline. Note both compression events (decisions arriving faster than tempo baseline supports) and extension events (decisions taking significantly longer than baseline).

Target threshold: Decision tempo should be stable and proportionate to the operational situation. Compression beyond what the orientation infrastructure supports indicates urgency pressure. Extension beyond what complexity warrants indicates ambiguity creation or deliberate stalling.

Diagnostic implication: Compression maps to Pathological Ally or pretext manipulation activity. Extension maps to Controlled Opposition Plant activity. Tempo whiplash, alternating rapid and stalled decisions, indicates active narrative destabilization. Stable tempo indicates healthy orientation.

Frequency: Continuous tracking against baseline.

5. Challenge Response Time

What it measures: How quickly and accurately leadership responds to legitimate questions, dissent, and contradiction.

How to measure it: Observe and record leadership responses to member challenges across meetings and operational discussions. Classify responses by type: fact-based clarification (accurate and prompt), deflection (redirecting without addressing the substance), defensiveness (treating the challenge as a threat rather than a question), or social consequence (the challenger experiences negative standing as a result of challenging).

Target threshold: All legitimate challenges should receive fact-based clarification. Any deflection, defensiveness, or social consequence attached to a legitimate challenge is a threshold event requiring immediate examination.

Diagnostic implication: Leadership behavior toward dissent is a more reliable indicator of infiltration risk and orientation health than any specific content being protected. A pattern of deflection or social consequence for challengers indicates that the narrative is being guarded rather than tested, which is the mechanism Chapter 11 identified as the beginning of orientation collapse.

Frequency: Continuous observation. Formalize the classification quarterly.

6. Drift Recovery Rate

What it measures: How quickly the group returns to documented baseline orientation after a detected drift event.

How to measure it: When a drift event is identified (language drift, alignment fragmentation, validation ratio shift, or archetype activity), measure the time from detection to demonstrated return to baseline across the affected metrics. Compare recovery time against the severity of the drift event and against prior recovery periods.

Target threshold: Recovery time should be proportionate to drift severity and should show improvement over successive cycles as the calibration architecture matures. A recovery time that is growing longer across successive events, or a drift event that does not produce recovery, indicates structural failure in the calibration architecture.

Diagnostic implication: Drift recovery rate is the single best indicator of the calibration architecture's operational health. A group that can detect and recover from drift quickly is a group that can sustain Mission Command through maintenance-level monitoring. A group whose recovery time is lengthening requires full GCP reactivation regardless of how stable it appeared before the drift event.

Frequency: Measured per drift event. Trend analysis quarterly.

Restoration Metrics

The following seven metrics are drawn from Vandergriff's five-action restoration sequence. They are designed for systems in active calibration and for assessing readiness to transition to

Mission Command. They function as the operational complement to the GCP metrics above.

7. Candor Index

What it measures: The percentage of after-action review participants who voluntarily admit a personal or unit shortcoming without prompting.

How to measure it: In each after-action review, track the number of participants who surface a personal or unit failure, error, or misalignment without being directly asked to do so. Divide by total participants to produce a percentage. Track across review cycles to establish baseline and trend.

Target threshold: Above seventy percent. Below fifty percent indicates a feedback environment in which accurate self-reporting is not culturally safe. A candor index trending downward across cycles is a high-priority signal regardless of absolute value.

Diagnostic implication: The candor index is the primary indicator of whether the feedback loop is functioning. A system with a low or declining candor index is sanitizing the data it feeds into its decision cycle. Mission Command cannot be safely employed in a system with a candor index below threshold because the shared understanding of intent and conditions that Mission Command depends on will be built on incomplete or distorted inputs.

Frequency: Every after-action review. Trend analysis monthly.

8. Initiative Events

What it measures: The frequency with which subordinates exercise disciplined initiative within commander's intent without seeking prior approval.

How to measure it: Track documented instances of subordinates making and executing decisions within the scope of established intent without returning to leadership for approval.

Distinguish between disciplined initiative (action within intent, appropriately scoped) and unauthorized departure from intent (action outside established parameters). Count only the former.

Target threshold: An increasing trend over time. The absolute number is less important than the direction of movement. A flat or declining trend in a system that has been running calibration protocols indicates that authority is not being genuinely distributed or that subordinates do not trust that acting within intent will be supported.

Diagnostic implication: Initiative events measure whether Mission Command is operationally real or merely declared. A system with low initiative events despite stated Mission Command employment is a system where subordinates have learned that the authority to act on intent is not genuinely available. That gap between doctrine and culture requires calibration before genuine decentralization can function.

Frequency: Tracked per operational or training cycle.

9. Truth-Teller Protection Rate

What it measures: The percentage of bad-news reports, challenge events, and dissenting assessments that result in positive or neutral consequences for the member who delivered them.

How to measure it: Track instances in which a member delivers bad news, challenges a flawed assumption, or surfaces a contradiction. Record the subsequent consequence for that member: positive (recognized, promoted, assigned increased responsibility), neutral (no change in standing), or negative (counseled, marginalized, assigned reduced responsibility, or socially penalized). The protection rate is the percentage of instances with positive or neutral outcomes.

Target threshold: One hundred percent. Any instance of negative consequence for a truth-teller is a threshold event that requires immediate leadership response. A single unaddressed negative consequence will suppress future truth-telling across the entire group, because members observe consequences and calibrate their behavior accordingly.

Diagnostic implication: Truth-teller protection rate is the structural test of whether the feedback culture the calibration phase is building is real. A system that claims to value candor but does not protect truth-tellers at one hundred percent has a culture that is declared rather than operational. Mission Command cannot function in that environment because the information it requires to operate will not surface.

Frequency: Tracked per event. Reviewed monthly.

10. Tempo Delta

What it measures: The difference between planned and actual time to achieve objectives in training or operational cycles, used as an indicator of decision cycle health.

How to measure it: In each training event or operational cycle, record the planned time to achieve each key objective and the actual time achieved. Calculate the delta for each objective. Track the trend across cycles. Consistent reduction in delta indicates improving decision cycle efficiency. Widening delta indicates decision cycle degradation.

Target threshold: A consistent reduction trend. Absolute values vary by operational context and cannot be standardized across environments. The diagnostic value is in the trend rather than the number.

Diagnostic implication: Tempo delta measures whether Mission Command is producing the decision speed advantage it is designed to produce. A widening delta in a system that has transitioned to Mission Command indicates that either the orientation foundation is degrading or the decentralization is not genuine. A consistently narrowing delta is evidence that the calibration phase produced what it was designed to produce.

Frequency: Per training event or operational cycle.

11. Trust Score

What it measures: The percentage of members who affirm, anonymously, that they trust their leadership to support them when they act within stated intent.

How to measure it: Administer an anonymous climate assessment with the question drawn from Vandergriff's restoration framework: "I trust my chain of command to support me when I act within intent." In civilian or non-military environments, substitute "leadership" for "chain of command" while preserving the rest of the phrasing. Use a five-point scale. Calculate the mean score and the percentage of respondents scoring four or above. Track across assessment cycles.

Target threshold: Mean score above four on a five-point scale, with at least seventy percent of respondents at four or above. A declining trust score in a system that has been running calibration protocols indicates that the structural trust architecture is not being experienced as genuine by the people operating within it.

Diagnostic implication: Trust score is the member-facing indicator of whether the calibration phase has produced its intended cultural effect. High structural trust metrics combined with a low trust score indicates a gap between the architecture as designed and the architecture as experienced. That gap requires investigation before Mission Command transition.

Frequency: Quarterly minimum. Monthly during active calibration.

12. Misalignment Admission Rate

What it measures: The frequency with which leadership publicly acknowledges a personal assumption error at dedicated calibration sessions.

How to measure it: Track the number of Map-Terrain Check sessions, or their equivalent in non-military environments, in which the senior leader present publicly identifies and names a specific personal misalignment: an assumption that proved wrong, a plan that did not match conditions, an assessment that required revision. The Map-Terrain Check is a structured weekly or recurring session in which leadership and members openly discuss one instance in which assumptions or plans failed to match reality, with the senior leader going first. Express misalignment admission rate as a count of sessions in which the senior leader admitted a personal error versus total sessions conducted.

Target threshold: Every session. A leader who cannot identify a personal misalignment in any given cycle is either operating without error (unlikely) or not looking (the diagnostic concern). The target is not perfection in performance but consistency in honest self-assessment.

Diagnostic implication: Misalignment admission rate is the leadership-level analog to the candor index. It measures whether the calibration discipline the leader is asking of subordinates is being modeled from the top. A leader with a low misalignment admission rate in a system with a high candor index has created a double standard that will eventually collapse the candor it has built. Consistent admission is both a diagnostic indicator and a cultural mechanism.

Frequency: Per session. Trend analysis monthly.

13. Follow-Through Rate

What it measures: The percentage of improvement commitments made in after-action reviews that are implemented within thirty days.

How to measure it: At the conclusion of each after-action review, record all commitments to specific changes, adjustments, or improvements, with assigned owners and thirty-day implementation targets. At the thirty-day mark, assess completion. Divide completed commitments by total commitments to produce the follow-through rate.

Target threshold: Above eighty percent. A follow-through rate below fifty percent indicates that the after-action review process is producing acknowledgment rather than change, which is the organizational equivalent of validation-seeking: the appearance of calibration without its substance.

Diagnostic implication: Follow-through rate is the test of whether the learning loop is closing. A system with high candor and low follow-through has identified its misalignments accurately but is not correcting them, which means the orientation data it is generating is not feeding back into improved performance. That gap between detection and correction is as dangerous as the gap between reality and the map, because it trains the system to surface problems without expecting solutions, which over time suppresses the candor that surfaced them.

Frequency: Thirty days after each after-action review. Trend analysis quarterly.

Using the Metrics Together

No single metric produces a complete picture. The diagnostic value of this framework comes from reading metrics in combination and tracking their movement over time.

A system with high candor index but declining trust score has a gap between what members are willing to say and what they believe leadership will do with it. A system with improving initiative events but widening tempo delta is decentralizing authority without improving decision speed, which indicates the orientation foundation is not supporting the decentralization. A system with a one-hundred-percent truth-teller protection rate but a rising language drift index is protecting its truth-tellers while losing its narrative, which means the archetype threat has shifted from the interpersonal to the cultural.

The combination of all thirteen metrics, assessed at regular intervals and compared against prior periods, is the operational expression of what orientation-based command looks like in practice: not a single assessment, not a gut feeling, but a continuous discipline of measuring the system's actual state against its documented baseline and acting on what the data says.

178

Acknowledgements

Kit Perez

I am beyond grateful to have worked with Don Vandergriff on this project. He brought decades of hard-won operational knowledge, a genuine openness to new ideas, and the kind of collaborative spirit that makes a co-author a partner rather than a problem. He was gracious, classy, and steady from the first conversation to the last page. I am honored to have built this with him.

Mark McGrath, co-host of the No Way Out podcast and The Whirl of Reorientation on Substack, deserves his own sentence for introducing me to Don, and several more for everything else. He helped me understand orientation at a depth I had not reached on my own, pulling me away from pop-OODA and toward Boyd's actual framework, and helping me name things I already understood but did not know Boyd had formalized. That grounding is the theoretical spine of GCP.

Bruce Gudmundsson's work on German military culture and stormtroop tactics gave me a precision about historical application that shaped how I understood what Mission Command was designed to produce. Sam Alaimo, Kyle Shepard, Mike Jones, Shane Copeland, Jonathan Kilburn, Steve Custer, MJ, Sarah Kernion, Ashton Richie, Shane Agee, Derith Shuler, and Rebecca Sulages span a wide range of disciplines and belief systems, and each of them contributed something to the thinking I brought to GCP over the years, whether they know it or not.

To the veterans who spoke honestly about serving under officers who believed they were running Mission Command but had never built the conditions it requires: your accounts are in this book, unnamed but present. You described the failure modes more precisely than any doctrine manual ever has. To the members of public and underground resistance organizations who kept getting waylaid by bad leaders or swallowed by resistance theater, and to the groups who did the ruthless work, demanding of their mission, their people, and themselves: you are the reason this doctrine exists in the form it does. You are also proof that it works.

To the many leaders who exemplified every failure mode in this book: GCP exists because of the cultures you created. I mean that without irony.

My husband Eric has been present for all of it. We met on a firing range. He looked at my rifle, told me with complete candor that it was bone dry, and has been exactly that kind of partner ever since: direct, engaged, and genuinely invested in making the ideas better. He has always seen the life we have built together as a true collaboration, and he is right. I would not be the thinker I am without him.

My son Alex challenges my thinking with a consistency and rigor that keeps me honest. He finds the angles I miss and surfaces the objections I have not considered. His wife Jordan has been a steady presence through the final stretch of this project. I am grateful for both of them.

Donald Vandergriff

This book is the product of a true partnership, and I begin by offering my deepest gratitude to my co-author, Kit Perez. Kit, your Grey Cell Protocols are the missing infrastructure that Mission Command has always needed but never had. Your brilliance in making explicit the hidden conditions that allow decentralized command to survive under pressure---especially in environments where formal authority is weak or absent---completed the framework I had spent decades searching for. Thank you for the intellectual generosity, the rigorous collaboration, and the shared conviction that orientation must be deliberately built before it can be exploited. This book is far richer and more useful because of you. I am honored to stand beside you as co-author and friend.

I am profoundly thankful to the mentors and thinkers who shaped my understanding of Mission Command and adaptive leadership over many years. William S. Lind and Dr. Bruce I. Gudmundsson opened my eyes to maneuver warfare and the historical roots of decentralized command. Mark McGrath, Brian "Ponch" Rivera of AGLX, and Johna Till Johnson of Nemertes, all who brought fresh, practical insight into applying these ideas in complex, high-stakes settings. Major PJ Tremblay (USMC ret.), LTC Asad "Genghis" Khan (USMC, Ret.), Colonel Douglas MacGregor (US Army retired Ph'D), Lt. Fred Leland (Walpole, MA Police, Ret.), Major Joe Labarbera (US Army ret.), MacKubin Owens (COL, USMC, Ret., PhD), Colonel Daniel Wittnam (USMC), and Séamus Na Gaeilge (US Army National Guard) each offered encouragement, critique, and real-world validation that strengthened every aspect of my work. Their collective wisdom and generosity continue to inspire me.

The work of the late Colonel John Boyd and his OODA Loop fundamentally changed how I see command under uncertainty. I am forever grateful to the Boyd acolytes who carried his ideas forward and shared them so openly: Franklin C. Spinney, Colonel G.I. Wilson (USMC, Ret.), COL Chet Richards (USAF, Ret., PhD), and the late Pierre Sprey. Their influence runs through every page of this book.

To my former cadets at Georgetown University who have carried Mission Command into their careers---both inside and outside the military---and adapted it to their own cultures: your success is my greatest reward. Watching you apply these principles with courage and results has been one of the true joys of my professional life.

I owe special thanks to my instructors at the USMC Amphibious Warfare School (now EWS), especially COL Fred Longoria (USMC ret), who encouraged, and continues to encourage, my early ideas and pushed me to think more deeply and rigorously.

To every law enforcement agency, military unit, government organization, and civilian/non-government group that has hosted my Outcomes Based Learning (OBL) and Mission Command workshops in recent years: thank you for your trust and, above all, for the moral courage to implement the best command culture. Your willingness to do the hard work of building the conditions has proven time and again that Mission Command is not only possible but transformative.

Finally, to my wife Lorraine: for twenty-eight years you have lived with my "other women"---the books, the travel, the late nights, and the relentless focus on reforming command culture. You have done far more than tolerate it. You have supported me, challenged me, and offered insights that made every idea sharper and more human. None of this would have been possible without your love, patience, and wisdom. This book, like everything else I have done, is as much yours as it is mine. I love you more than words can say.

Other Works by Donald E. Vandergriff

Books

Spirit, Blood and Treasure: The American Cost of Battle in the 21st Century, ed. Novato, CA: Presidio Press, 2001.

The Path to Victory: America's Army and the Revolution in Human Affairs. Novato, CA: Presidio Press, 2002.

Raising the Bar: Creating and Nurturing Adaptability to Deal with the Changing Face of War. Washington, DC: Center for Defense Information, 2006.

Manning the Future Legions of the United States: Finding and Developing Tomorrow's Centurions. Westport, CT: Praeger Security International, 2008.

Adopting Mission Command: Developing Leaders for a Superior Command Culture. Annapolis, MD: Naval Institute Press, 2019.

The Outcomes Based Learning Professional Handbook: Building Effective Leaders and Teams. Petersburg, VA: Special Tactics Institute and University, 2022.

Anthologies Edited

Mission Command: The Who, What, Where, When and Why --- An Anthology, ed. with Stephen Webber. North Charleston, SC: CreateSpace, 2017.

Mission Command II: The Who, What, Where, When and Why --- An Anthology, ed. with Stephen Webber. North Charleston, SC: CreateSpace, 2019.

The New Maneuver Warfare Handbook, with William S. Lind. Petersburg, VA: Special Tactics Institute and University, 2023.

Chapters in Edited Volumes

"The Culture Wars." In Digital Wars: A View from the Front Lines, ed. Robert L. Bateman III. Novato, CA: Presidio Press, 1999.

Fiction

Reforging the Sword. North Charleston, SC: KDP, 2026.

Operation Desert Fang. Book 2 in the Reforging the Sword Series. North Charleston, SC: KDP, 2026.

Substack

Don Vandergriff. donaldvandergriff.substack.com.

Vandergriff has authored, edited, or contributed chapters and appendices to approximately thirteen books in total, including additional forewords and contributions to volumes on maneuver warfare and leader development. He has authored more than 150 articles on military reform, leader development, and Mission Command. Several of his books have been translated into Ukrainian, Taiwanese, and German.

Other Works by Kit Perez

Books

The Mindset of Resistance: Saying No to Tyranny in an Effective Way. 2021.

Basics of Resistance: The Practical Freedomista, Book I, with Claire Wolfe. 2018.

Substack

The Shepard Scale. shepardscale.substack.com.